CONTRACTS AND INTERNATIONAL PROJECT MANAGEMENT

Contracts and International Project Management

DAVID G. CARMICHAEL
The University of New South Wales, Sydney, Australia

A.A. BALKEMA / ROTTERDAM / BROOKFIELD / 2000

Published by
A.A.Balkema, P.O.Box 1675, 3000 BR Rotterdam, Netherlands
Fax: +31.10.4135947; E-mail: balkema@balkema.nl; Internet site: http://www.balkema.nl

A.A.Balkema Publishers, Old Post Road, Brookfield, VT 05036-9704, USA
Fax: 802.276.3837; E-mail: info@ashgate.com

ISBN 90 5809 324 7 hardbound edition
ISBN 90 5809 333 6 student paper edition

Contents

ABOUT THE AUTHOR   VIII

1 INTRODUCTION   1

2 IN-HOUSE VERSUS OUTSOURCING   5
   2.1 Introduction   5
   2.2 In-house approach   5
   2.3 Outsourcing – Contract approach   6
   2.4 Public sector and private sector work   9
   2.5 Case studies   15
        2.5.1 Case study – Road construction   15
        2.5.2 Case study – Contractor decisions on outsourcing   16
        2.5.3 Case study – Outsourcing coal mining   18
        2.5.4 Case study – Local government practices   19
   2.6 Exercises   22
   References   24
   Bibliography   24

3 CONTRACT (PAYMENT) TYPES   25
   3.1 Introduction   25
   3.2 Outline   25
   3.3 Fixed price contracts   27
        3.3.1 Lump sum contracts   30
        3.3.2 Schedule of rates contracts   33
        3.3.3 Guaranteed maximum price (GMP) contracts   40
   3.4 Prime cost contracts   40
   3.5 Incentives/disincentives   55
   3.6 Choice of contract (payment) type   62
   3.7 Case studies   65
        3.7.1 Case study – Sewer reticulation project   65
        3.7.2 Case study – Substation dismantling project   67
        3.7.3 Case study – Building excavation contract   69
        3.7.4 Case study – Redevelopment of a large city park   70
        3.7.5 Case study – Construction of a shopping centre   73
        3.7.6 Case study – A contractor's business   74

	3.7.7	Case study – Local government – flooding and drainage	76
3.8	Exercises		77
References			82

4 DELIVERY METHODS   83
4.1   Introduction   83
4.2   Outsourcing options   87
    4.2.1   Case study – Contaminated sites   95
4.3   Traditional   101
    4.3.1   Case study – House construction   105
    4.3.2   Case study – Road construction   106
4.4   Design-and-construction   107
    4.4.1   Detail design-and-construction   111
    4.4.2   Design-and-construction   113
    4.4.3   Case study – Hospital project   117
    4.4.4   Case study – Road building projects   119
    4.4.5   Managing contractor   120
4.5   Novation   121
    4.5.1   Case study – One company's experiences in the building industry   124
    4.5.2   Case study – Stormwater retention basin   127
4.6   Design-construct-and-maintain   129
4.7   Project management   131
    4.7.1   Alternative configurations   135
    4.7.2   Case study – Can-making   138
4.8   Design and management   141
4.9   Construction management   141
    4.9.1   Alternative configurations   148
    4.9.2   Case study – Instrumentation project   150
    4.9.3   Case study – Owner-builder   153
4.10 Concessional methods   154
    4.10.1 Case study – Water treatment project/plant   160
    4.10.2 Case study – Housing village   162
4.11 Fast-track   164
    4.11.1 Case study – Infrastructure project   166
    4.11.2 Case study – Internal upgrade project   172
4.12 Single or multiple contracts   175
    4.12.1 Bundling   176
    4.12.2 Staging   178
4.13 Performance of different delivery methods   179
4.14 Exercises   181
References   184
Bibliography   185

5 INTERNATIONAL CASE STUDIES   186
5.1   Outline   186
5.2   Case studies   186

5.2.1   Case study – Port development, Asia   186
5.2.2   Case study – Hydroelectric power station project, Asia   190
5.2.3   Case study – Hydroelectric project, Asia   193
5.2.4   Case study – Condominium project, Thailand   196
5.2.5   Case study – Development of public facilities, buildings, Indonesia   199
5.2.6   Case study – Airport extension, Cambodia   201
5.2.7   Case study – B&T delivery, toll road, Indonesia   204
5.2.8   Case study – Lump sum contract, discharge, civil work, Indonesia   205

SUBJECT INDEX   207

About the author

David G. Carmichael is a graduate of The University of Sydney (B.E., M.Eng.Sc.) and The University of Canterbury (Ph.D.) and is a Fellow of The Institution of Engineers, Australia, a Member of The American Society of Civil Engineers, a Graded Arbitrator with The Institute of Arbitrators, Australia and a trained mediator. He is currently a Consulting Engineer and Professor of Civil Engineering, and former Head of the Department of Engineering Construction and Management at The University of New South Wales.

He has acted as a consultant, teacher and researcher in a wide range of engineering and management fields, with current strong interests in all phases of project management, construction management and dispute resolution. Major consultancies have included the structural design and analysis of civil and building structures; the planning and programming of engineering projects; the administration and control of infrastructure projects and contracts; and various mining, construction and building related work.

He is the author and editor of 15 books and over 65 papers in structural and construction engineering, and construction and project management.

CHAPTER 1

Introduction

Outline

This book focuses on a number of important contractual aspects of project management and provides this in an international context:

- Whether to use an organisation's own resources, or use or contract with resources outside the organisation (Chapter 2 – In-house versus outsourcing).
- The choice of method by which contractors and consultants are paid (Chapter 3 – Contract (payment) types).
- The choice of contractual relationships to use between all the parties to a project (Chapter 4 – Delivery methods).

International case studies illustrate their application in practice (Chapter 5 – International case studies).

Procurement

Procurement refers to the practices involved in engaging resources from outside (and also possibly within) an organisation to:

- Do work (construct, fabricate, manufacture, ...).
- Provide services (management, consulting, design, research and development, ...).
- Supply materials and products.
- Design, manufacture and/or supply equipment.

The broad components of procurement can be seen in Figure 1.1. The shaded boxes represent the topics covered in this book. Attention is focussed on project work and services. Issues related to purchasing or procuring materials and equipment are largely excluded, though they obviously overlap. Tendering and conditions of contract are mentioned as they relate to the other topics, but are also largely excluded.

Contract (payment) type

In most commercial contracts, one party pays for the work, services, materials, ... of the other party. There are a number of payment options that owners can consider, leading to the various contract (payment) types.

Different contract (payment) types offer different incentives to the contractor or consultant and influence the cost, delivery time and performance on projects. 'Payment' here

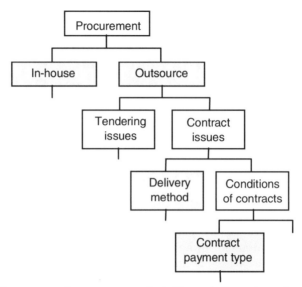

Figure 1.1. Components of procurement – shaded boxes indicate the topics covered in this book.

refers to the overall way the contractor or consultant is reimbursed or paid for its role in the contract.

On any given project with many contracts (between the various players or stakeholders – owner, consultants, contractors, subcontractors, suppliers, ...), there may be a whole range of contract (payment) types, each chosen to reflect the circumstances particular to the nature of the work.

Delivery method

The delivery method allocates the different project functions (like-activities, or sub-projects – design, management, construction, ...) to the various project participants (typically at the organisation level, rather than the individual level); it establishes the roles and interrelationships of the project participants, and influences management practices. It can involve issues of timing, but excludes specifics of contractual details (such as conditions of contract including contract payment type). The usage of the term 'delivery method' applies at the organisation level and not lower.

The terms *procurement, contact,* or *delivery* in conjunction with *system, strategy, method, approach* or *arrangement* and similar may be used synonymously and interchangeably, and in different senses by different people (Fig. 1.2). All relate in some way to contractual and/or non-contractual means of getting project activities done. The reader needs, therefore, to be careful when looking at other documents, as to the usage of terminology in these documents.

Some people precede this terminology with the word 'project', but this is considered too ambitious, because generally it is only a few aspects of a project which are being referred to and not the whole project (though the writer acknowledges the definition of a 'project' as sufficiently flexible to allow such ambitious terminology).

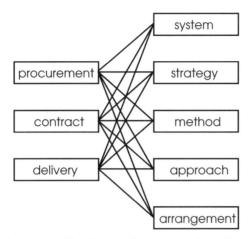

Figure 1.2. Various terminology combinations used synonymously and interchangeably and in different senses by different people.

The term 'delivery method' has been adopted in this book. The word 'method' is used in the sense of 'a way of proceeding or doing something', 'orderliness of thought action', and 'arrangement of work'. Figure 1.1 helps to clarify the usage of terminology in this book.

Horses for courses

It is usually described as 'horses for courses' (that is, backing the horse that performs well on a given racetrack) when selecting the vehicle to oversee the contract work. There are many vehicles resulting from the various combinations of contract (payment) types and delivery methods. Some people and organisations have a preference for one vehicle over another, perhaps because of familiarity. Others steadfastly remark that a certain vehicle is better than another; most of these viewpoints are based on anecdotal evidence rather than any real objective evidence.

Terminology

Terms used interchangeably by writers include:
• Owner, client, principal, employer, developer, proprietor, purchaser.
• Contractor, builder.
• Owner's representative, superintendent, architect, engineer.
Generally, the first term in each of these lists is the one adopted in this book.

Owner-contractor

The book is written largely in terms of an owner-contractor relationship. The notes generally apply to the contractor-subcontractor relationship, and dealings with consultants and suppliers as well. Only where there is a specific need to refer to consultants or subcontractors, rather than the more general term contractor, is such specific terminology used.

Gender comment

Commonly references use the masculine 'he', 'him', 'his' or 'man' when referring to project personnel possibly because the majority of project personnel have historically been male. However such references should be read as non-gender specific. Project management is not an exclusive male domain.

Acknowledgement

The book contains numerous case studies contributed by as many people. Their contributions are gratefully acknowledged.

CHAPTER 2

In-house versus outsourcing

2.1 INTRODUCTION

Options

Broadly there are two options available to any owner to get project work done. The work can be done (Fig. 2.1):
1. *In-house* (referred to as *direct control*, direct labour, departmental work, or other terms).
2. Using outside resources – *outsourcing* (i.e. by *contract, contracting-out,* also referred to as farming-out).

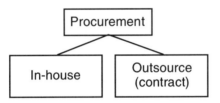

Figure 2.1. Procurement options.

(Note that the terms *day labour* and *daywork* are used to refer to a contractor's staff being employed on an owner's needs basis and the contractor being reimbursed by the owner for the staff's time, together with some supplementary margin, that is effectively on a cost-plus basis. However, occasionally the terms are used more widely to refer to labour hired directly by the owner, and are sometimes used synonymously with 'in-house'.)

It is also possible to have a *combination* of in-house and outsourcing. Examples include: outsourced supply of crushed stone, but transported using in-house vehicles and drivers; mine development and expansion, exploration, drilling, mine design and maintenance outsourced, but ore extraction in-house.

2.2 IN-HOUSE APPROACH

An in-house approach refers to the owner undertaking the work using its own resources, and capabilities internal to the owner's organisation. As a result, the owner can achieve any desired standard of work and rate of progress. These standards therefore need not be laid down explicitly as there is no need to convey the ideas and subconscious requirements of the owner to another person or firm.

Terms used synonymously with 'in-house' include 'direct control', 'direct labour', 'departmental work' and similar (and sometimes 'day labour' and 'daywork').

Work undertaken using in-house resources has often been criticised on the grounds that the approach is inefficient. However, this is not necessarily true, particularly where management and the organisation's culture are attuned to good practices. Efficiency using in-house resources can be achieved where, amongst other things, the owner carries out a continuous large volume of specialised work and hence can employ specialised labour, plant and techniques on a long term basis. Maintenance, refurbishment, retrofitting and rehabilitation work can be carried out efficiently using in-house procurement. Its use in 'greenfield' work will depend on individual circumstances.

The main advantages and disadvantages of in-house procurement are as follows.

Advantages

- Greater control by owner; possible for the owner to react to changed circumstances.
- Flexibility in timing and funding.
- Ability to react quickly by redeployment of resources; short lead time requirements; emergencies and urgent work.
- Useful where there is a limited availability of contractors, contractors don't want the work, or available contractors are unsuitable.
- Useful where the market price of work is being governed by a limited availability of contractors.
- Suitable for very small jobs; on-going specialised work; where intellectual property is involved, for example research and development work; projects with ill-defined scope, for example experimental work; projects with undeveloped documentation.
- Develops in-house expertise.
- Avoids potential collusion on prices between private providers.
- Useful for interface work between contracts.
- Can be supplemented with hired equipment and people.

Disadvantages

- Not subject to competition.
- Subject to resourcing constraints; multiple simultaneous projects may not be possible.
- Requires on-going work; not suitable for one-off or irregular work.
- Owner commitment and requirements not necessarily clearly defined.

2.3 OUTSOURCING – CONTRACT APPROACH

Under a contract approach, the project owner engages a specialist firm or individuals – capabilities external to the owner's organisation – to do the required work and, under the right circumstances, this approach can be more efficient than the in-house approach.

If the project owner does not have a long term program of similar work, it is not appropriate for the owner to establish or equip a group to carry out a limited program. Reasons for not doing so include lack of experience in certain phases of the work and the problem of establishing and disbanding a facility or group over a short period of time. The limited volume of work involved would also result in inefficiencies.

Under these situations, an owner would possibly be better off using a contract approach for undertaking the work. By doing so, the owner can draw on the expertise and resources of a specialist firm in the area and, providing fair competition exists between contractors, the owner will obtain an economically efficient project.

Because the contractor is undertaking work for somebody else, it is firstly very necessary for the owner to state exactly what is wanted in order that the owner's requirements are fulfilled and the contractor is left in no doubt as to what is wanted. As a result, full contract drawings and specifications may need to be provided by the owner beforehand.

Because of competitive bidding between contractors for work, profit margins may be small. Hence contractors may take opportunities to cut costs. This they are entitled to do provided that the work produced is in accordance with the drawings and specifications. This gives rise to the problem of supervision and contract administration by the owner, and quality assurance.

If the owner is not experienced in this type of work it may be advantageous to employ an agent to supervise and manage the contract on behalf of the owner.

Contracting out work is now an accepted practice in areas such infrastructure development, mining, computing and so on. Core competencies might be retained in-house, with everything else outsourced. The specialist expertise of contractors and consultants is tapped on a needs basis. Contractors and consultants are significant players in many industries.

Owners may use contractors and consultants as a flexible alternative to increasing permanent staff, augmenting personnel shortages, or to accommodate growing programs of work. Increasing permanent staff for temporary increases in work brings with it traumas associated with personnel layoffs, and delays associated with hiring.

The main advantages and disadvantages of going the contract route are as follows.

Advantages

- Owner commitment and requirements clearly defined.
- Ability to use specialist expertise and resources not available within the owner's organisation.
- Permits multiple simultaneous projects by providing access to quantities of resources.
- Permits a one-off, short timeframe project, without hiring/firing.
- Permits an early start to projects; contractor may be able to mobilise resources quickly.
- Suitable for irregular work; resource requirement peaks and troughs satisfied by contractor mobilising resources as appropriate.
- Introduces competition.
- Permits risk transfer to the contractor; e.g. responsibility for industrial relations; e.g. final project cost.
- Avoids capital expenditure up-front; may improve cash flow.

- Enables an organisation to grow (in terms of work volume) quickly without taking on extra fixed costs.
- Permits the owner's organisation to focus on core competencies.

Disadvantages

- Diminished control by the owner.
- Less flexibility to make changes at minimal cost.
- Resources are consumed in the production and administration of a contract.
- The cost of dispute avoidance and potential disputes.

Outsourcing

Outsourcing of parts of a project and the completed facility may be with respect to, for example:
- Finance,
- Design,
- Supply,
- Management,
- Construction (building) / fabrication / manufacturing,
- Operation,
- Maintenance,
- 'Owning' (over different franchise periods or indefinitely); leasing, and
- Combinations of these.

Influence of outsourcing

Outsourcing might be thought of as converting project work into 'arm's-length' relationships with those doing the work. Specialist firms have evolved to perform work that others don't wish to perform. Outsourcing permits an organisation to grow (in terms of work volume) quickly without taking on extra fixed costs and with reduced risk.

Extreme forms of outsourcing lead to an organisation with essentially no employees. The organisation becomes devoid of expertise. The organisation becomes dependent on the practices of contractors, particularly with respect to issues such as quality. Whether total, partial or no outsourcing is practised in an organisation, and the resulting efficiencies and costs, will depend on the situation, though some people approach it in a faddish fashion.

It is suggested that outsourcing, approached in terms of developing long-term relationships and using long-term strategies, rather than being based on short-term savings in overheads, may lead to larger savings. Partnering ideas are similar. However, some people feel uncomfortable with long term relationships, and not testing the market periodically for competitiveness; their preference is to have contractors continuously compete for work.

Outsourcing, to some people, is a way of washing their hands of a problem. However, the contractual and business relationships with the contractor need to be managed, and a hands-off approach could be expected not to work.

There is a view that if an owner gives an organisation work, and the organisation subsequently outsources some or all of that work, then that organisation's relationship with the owner is weakened. This is the typical situation contractors face when engaging subcontractors; some contractors do no work other than manage the subcontractors. If the subcontractors don't perform to expectation, owner blame may be directed at either the contractor or the subcontractor.

Historically, and particularly in the public sector, industrial unions have fought the introduction and movement towards outsourcing. Employers, on the other hand, have used economic arguments to introduce outsourcing and to eliminate work practices and worker entitlements with which they did not agree.

Contractor and consultant selection

Preference is for contractors to be selected via a prequalification process.

The preferred way of selecting consultants is through interview, where their qualifications, their ability to resolve potential problems, their people-relationship skills, and their knowledge of the project work can be evaluated.

There may be difficulty in establishing the scope of work of consultants, and in estimating their costs. The costs will vary with the type and size of the project, as well as the expectations of the owner.

2.4 PUBLIC SECTOR AND PRIVATE SECTOR WORK

Public sector work

Historically, public-sector bodies have favoured the in-house procurement option for much of their work. But the trend appears to be now away from this, promoted by a number of reasons:
- Government regulations encouraging or forcing contracting-out.
- The requirement for greater accountability.
- Efficiencies (cost savings) through the introduction of competition.
- An economic and moral obligation on the public sector to be efficient in its use of community funds.

The debate, during its early stages, over contracting out public sector services was considered 'controversial' (Fraser, 1992). Some people regarded it as the solution to the problems facing the public sector; others regarded it as creating more problems than it solved. Contracting out public sector services is regarded as a form of *privatisation*, which has political undertones. Contracting out of some services has always existed; however it was the trend to contracting out core services which intensified the debate.

Arguments in favour of contracting out public sector work

Proponents of contracting out public sector work list the following in their favour:
- Improved economy and productivity through competition and market forces, and decreased interference from trade unions (e.g. industry awards with service agreements

protecting incumbent employees; inflexible work environment; rigid work practices; limits on productivity; wages and salaries not linked to productivity; etc.) and government regulations; the benefits extend to work awarded to in-house groups if they are subject to the same competition.

- There is a widespread perception that the public sector is less efficient than the private sector because of interference from trade unions, government regulations and public sector work culture.
- There is a perception that some people in the public sector deliberately sabotage projects that have been contracted out to the private sector, in order to show privatisation in a bad light.
- Cost savings (up to 20%), with no deterioration in quality of service.
- The benefits of contracting-out are claimed to have been demonstrated in other countries.
- Arguments against contracting-out are often assertions without rigorous and thorough analysis to support them; there are flaws in the research methodology (e.g. limited or selective samples in quantity and time; selective documentation; partisan researcher).
- Benefits to the public sector may be better obtained through improved service delivery methods, rather than any inherent private sector – public sector difference.
- Smaller government, and privatisation are desirable.

Arguments against contracting out public sector work

Proponents against contracting out public sector work list the following in their favour:
- Arguments in favour of contracting-out are often assertions without rigorous and thorough analysis to support them; there are flaws in the research methodology (e.g. direct cost alone may be considered without regard for example to the 'hidden' costs or quality; limited or selective samples in quantity and time; data may not distinguish between labour, equipment and other costs; selective documentation; partisan researcher).
- Claims of improved economy and productivity through competition and market forces, and claims of public sector inefficiencies cannot be substantiated; there are inefficiencies within the private sector – to say that the private sector is always more efficient than the public sector cannot be substantiated.
- There is the potential for private providers to collude over the prices tendered.
- Jobs are lost (though some may transfer to the contractor organisation), employee conditions reduce and private monopolies develop.
- The benefits of contracting-out, claimed to have been demonstrated in other countries, are not without question; for example in some instances costs have increased, particularly where there have been contract failures.
- The cost savings quoted appear to only occur when high standard, successful contract management practices are followed, and include such things as prequalification of tenderers, incentives, penalties, regular inspections, fixed price contract type, resolution of disputes by negotiation etc; that is, precautionary practices are necessary on the part of the owner to ensure the private sector performs; it follows that the private sector does not perform to stated efficiencies unless the owner is vigilant and interventionist; it also follows that some of the cost savings are consumed by additional owner activities.

- Cost savings are offset by increased contract administration costs and other 'hidden' costs.
- Technical expertise is lost, or diluted (people shift to private companies) for the sake of initial apparent efficiency gains, based solely on short- to medium-term financial outcomes. Potential training grounds for new professionals disappear. Expertise necessary for planning infrastructure disappears.
- Against smaller government and privatisation – government exists in the economy to deal with market failure, corporate concentration, wealth distribution and social equity; so-called 'inefficiencies' in government are there to meet community service obligations, protect the environment etc.
- There may be social equity employment problems on going to the private sector. Private contractors are generally guided by the profit motive first and social responsibility issues second.
- Smaller rural communities are reluctant to give work to contractors from distant large cities, because of losing local employment, resources and expertise.

Activities suitable for contracting-out

It is not applicable to all services
Many services and activities are not suitable candidates for competitive tendering, for example, strategic functions and those where there would be unacceptable risks. The identification of activities which are suitable and those which are not is a matter for each organisation.

The activities which are most likely to offer scope for competitive tendering can be broadly defined as:
- *Those where the market place has clearly established the relevant capacity.*
- *Those which are not part of the core business of the agency.*
- *New activities where the option of contracting-out should be appraised automatically before the work is absorbed into the agency program.*
- *Those that are relatively discrete from other activities.*
- *Those subject to wide fluctuations in workload requiring considerable adjustments to staffing.*
- *Those which are part of a quickly changing market and where it is costly to recruit, train and retain staff.*

(Office of Public Management, 1991)

The general feeling is that it is desirable to have some public sector work contracted out, but that all work is not suitable to be contracted out. The exercise, then, is to determine which work is suitable for contracting-out, and which is not suitable. An informed decision is necessary. There are a number of reported cases of the public sector returning to doing the work in-house because of concerns over quality of service, additional unexpected costs etc.

Private sector outsourcing

Many of the issues facing the public sector, on whether to contract-out, apply to the private sector, particularly with the present-day public sector being asked to be accountable for its expenditure, and the public's expectation of 'value for money'. Leading issues relate to:

- Core competencies.
- Expertise.
- Financial strength.
- Economics.
- Spreading risk.
- Immediacy.

Core competencies

Consideration is given as to whether the work is a core activity of the organisation, or whether it is peripheral. The organisation needs to ask itself what is its core business. Activities outside the core business are candidates for outsourcing. Core activities are also candidates, but less likely ones because, in the long term, the organisation has no competitive advantage over other organisations who could employ the same contractors.

Expertise

Organisations have expertise in some areas and not others. It is rare for organisations to be uniformly expert across their whole range of functions. Outsourcing in the areas where expertise and capabilities are lacking enables the organisation to focus and concentrate its resources on its strengths. Contractors are able to provide breadth and depth in areas deficient in resources.

Financial strength

Each organisation will make decisions, based on financial strength, as to what is acceptable capital investment, operating costs and cash flow, allowing for taxation, investor and other considerations. The outsourcing option assists this decision making.

Economics

Even though an organisation may have the resources and capacity to do the work, outsourcing may nevertheless be a more economical alternative.

Spreading risk

Outsourcing enables risk to be transferred to the contractor, albeit more than likely at a cost to the organisation. This is done through appropriate choice of conditions of contract, payment type and delivery method. For example, responsibility for industrial relations issues can be transferred to a contractor. The organisation outsources risk to a level at which it feels comfortable it is prepared to accept. Outsourcing can be seen as one aspect of risk management on projects.

Immediacy

Where results are required in a (short) timeframe unable to be met by in-house resources, contractors can provide ready resources that may be capable of doing the work at short notice.

Case examples

Gold Mine, Western Australia

The gold mine was typical of WA gold mines consisting of numerous small deposits with low tonnage in relatively close proximity to each other. The possibility of turning this deposit into a mine drove its purchase, however the known resource at the time made it a stretch to justify developing the mine. Despite the prospectivity, corporately the owner had to ask itself, 'How much capital is it willing to risk on the basis of upside potential?' The answer to this rhetorical question was, 'Not much'. The owner found it difficult to commit owners' equity on something other than hard numbers and had to work very hard to turn this deposit into a mine.

Contractor mining was chosen for the property. The conditions were favourable for this decision. There was a large pool of contractor expertise available. Adopting contractor mining reduced the capital risk of the project, and provided economics which met corporate criteria.

The reserves substantially increased after start-up as a consequence of continued exploration and purchase of deposits in the area.

Adopting contractor mining reduced the owner's risk on this project. Of course this was somewhat of a two-edged sword as it also fixed the upside reward if the reserve base was significantly expanded. Finding more ore resolves the ownership cost issue. When this happens, the question then becomes, 'Are the contractor's efficiencies sufficient to offset its profit and overheads?' In short, 'Can the contractor mine at a lower cash cost than the owner?' As the owner sees mining as a core competency, and fundamental to its business, generally it would believe it can mine at a cash cost less than a contractor because it can eliminate the contractor's profit and overheads. As a consequence, for the owner, long-life projects inevitably drive the economics towards owner-operated mining.

It is easy to ask the question with the benefit of hindsight, 'Was adopting contractor mining a correct decision?' The answer has to be an unequivocal Yes. The owner made a business decision at the time, based on the available information, which fitted its corporate risk profile, and turned a resource into a mine. It is hard to be retrospectively critical about a decision based on known data and which brings value to the shareholder.

Copper/gold mine, Queensland

The mine life currently stands at 11 years as an underground operation.

Owner-operated mining was selected for the underground operation. This decision was a relatively easy one as mining was a core competency of the owner and it was possible to justify the purchase of mining equipment given the long mine life. However, contractor mining, was selected to carry out the initial underground development.

Contractor mining was also chosen to develop an open cut that provided early mill feed while underground development was progressing. The open cut also provided a convenient entrance for a production decline.

The mine is a good example of purchasing expertise to address a short timeframe problem. There was never any real discussion on whether the owner's equipment should mine the open cut. It was just too short a timeframe and required a different expertise (open cut mining) than what was needed for the long term.

(Based on: A.L. Hills, Use of Contractors at Placer Dome, Contract Operators' Conference, 28-29 October, 1966, Kalgoorlie, The Australasian Institute of Mining and Metallurgy.)

Cost savings

It is unclear what the actual cost savings are in going from in-house procurement to out-sourcing or contracting-out. Also, concentrating on price alone, as a comparison between in-house and outsource, is not appropriate. Some writers suggest in the region of a 20% cost saving. However this is debatable, and may only be in the region of 5% (and some people even suggesting 0%), for a number of reasons (not usually costed in a comparison):

- The cost savings quoted appear to only occur when high standard, successful contract management practices are followed.
- Contractors may offer a lower price by compromising standards or reducing the wages and conditions of employees; quality of service and accountability may be issues.
- The introduction of the tendering process itself adds costs, some writers suggesting in the region of 5% to 10% of the contract price. There is an administration cost attached to supervising contractors and inspecting their work.
- There is the potential for disputes, and their resolution or avoidance is a cost. The cost is non-value adding to the project.
- The value of 'after-sales service' obtainable by having on hand the people who did the work, after the work is complete, tends to be downplayed. At the completion of the work, contractors depart, unless maintenance is part of the contract.
- With a contract, the knowledge and skills gained from doing the work are lost to the organisation on completion of the work. No long term knowledge and skills can be developed and continual improvement processes implemented within an organisation, except for owner administration practices.
- Contractors incorporate a profit/margin/mark-up into their price; no profit component exists in in-house work.
- On changing over to contract, there may be the 'hidden' costs associated with staff redundancies, staff retraining, staff redeployment, changes in pay and employment, and equipment non-utilisation, including senior management time in closing down activities and dealing with the human resource issues.
- The best way to maintain cost savings through contracting-out may be to avoid a private monopoly evolving, through the public sector continuing to tender for work. This also prevents potential collusion on contract prices between private firms.
- Where the public-sector body also promotes its own resources as a candidate tenderer in order to demonstrate its competitiveness, there are additional administration costs keeping the tendering process and tender evaluation separate from the in-house tenderer.
- On first opting to contract-out by a public sector body, contractors may undercut the in-house tenderer in order to win the work, eliminate the in-house competition and possibly gain a monopoly on future work. Future work could be expected to be tendered at a higher price.

Competitiveness

In attempting to demonstrate their competitiveness, many public-sector bodies allow themselves to tender competitively against outside contractors. The difficulty that has been experienced with this in some cases is that contractors believe that there is not a level playing field. Specifically the way the public-sector's tender price is put together is sometimes questioned.

When [an in-house-work] organisation puts in an estimate for work to be carried out it should, in addition to the direct costs of labour, plant and materials, include the following costs:
- *Plant – All costs of financing the heavy capital expenditure involved and of operating and maintaining the plant, with a proper allowance for depreciation.*
- *Transport – All costs of financing, operating and maintaining its own transport, with a proper allowance for depreciation. Where the local authority's transport is used, the full cost of this should be included.*
- *Overheads – A correct allocation of overhead costs such as engineers, surveyors, clerks, administration, audit, accommodation, heating and lighting used by the ... organisation.*
- *All labour on costs – Including extra items such as annual leave, long service leave, payroll tax, worker's compensation charges, fares and travel allowances.*
- *Interest payments – All interest payments on the money required to finance the job.*
- *The maintenance period – An allowance for the cost of correcting any defects that may appear during the maintenance period, when the [facility] is under actual operating conditions.*

(AFCC, 1972)

One of the main advantages of in-house, namely its flexibility, can also be used as an argument against in-house and in favour of contracting-out. In particular, the following issues are raised:
- Were an estimate and specification prepared beforehand?
- If so, was the work done to budget and specification?
- Who is responsible for defects and maintenance?

By contracting-out, instead of in-house, it is claimed that all these matters are accounted for. In-house accountability in these matters may not be present.

2.5 CASE STUDIES

2.5.1 CASE STUDY – ROAD CONSTRUCTION

A road authority contains internal project management, construction and consulting divisions. The engagement of these divisions (by the parent organisation) is not done by formal contracts.

On some projects, an external project manager, and/or an external contractor, and/or an external consultant are engaged.

The project management division appoints a project manager (an employee) for each project, or number of projects, depending the size and complexity of the projects. In the case of contracting out the construction, the project management division prepares all documents before letting the contract. However, when the construction division undertakes the project, very often, documents are only partially developed; detailed design and document preparation are carried out while the construction is progressing.

The project manager prepares a master program, quality assurance program and budget for each project. Continuous review is necessary. A single point contact is maintained between the project management division and the construction division for the delivery of the project. The project manager is expected to be familiar with the issues related to industrial relations and safety in the workplace. The project manager regularly communicates with the parent organisation and the community.

EXERCISE

1. Engaging the project management division has both advantages and disadvantages. This gives greater flexibility to the parent organisation in engaging suitable/qualified/trained project managers for the projects. The employee project managers are well trained and well informed about the parent organisation's interests and responsibilities. However, there is the possibility for someone to take things for granted because an employee of the project management division is also an employee of the parent organisation. How else besides commercialisation and continuous improvement can these issues be addressed?

2. In the case of using the construction division, again, the employees of the construction division are employees of the parent organisation. This has various advantages and disadvantages. It gives the parent organisation the flexibility of commencing works while the documentation is still being carried out. This may lead to poor construction programming and inefficient use of resources. Also, there is a possibility of project controls such as on time and cost not being strictly adhered to. In some cases, resource optimisation cannot effectively be done due to the organisational culture (work gangs tend to stay together irrespective of the size of the work). What else besides work culture is influencing this downside comment?

3. Quality is always guaranteed due to the fact that the authority employees normally do not compromise the quality for the sake of reducing cost. Also, since the project management division and the construction division are both part of the same organisation, good relationships are maintained and project goals are achieved through smooth progress with less paperwork.

   When external consultants are engaged to manage the projects which are constructed by the construction division there is the potential for conflict. What is the basis of this conflict?

2.5.2 CASE STUDY – CONTRACTOR DECISIONS ON OUTSOURCING

Construction of an access road and a bridge to a new residential development

Concrete work

The residential subdivision was the first in a new area zoned for development. The only all-weather access to the site and the only access for connection of the services of water, electricity and telephones planned for the area were through the construction of a bridge. Delays were encountered with council approval of the development. Most of the blocks of land in the previous stages were sold, resulting in the developer being in need of land to sell. Subsequently, pressure was applied to the contractor to construct the bridge, road and subdivision as fast as possible so that the blocks could be sold.

In endeavouring to achieve the fastest possible construction time on the bridge (which was the critical section of the work) it required the concrete work to progress rapidly.

To achieve this, the contractor looked at different procurement approaches for concrete work. Past experience of concrete subcontractors was that they were generally unreliable when it came to adopting the contractor's program of works. For this reason, the contractor considered the execution of the concrete work using directly-hired labour (and in-house supervision). The advantages of this approach were seen to be:

- Greater control of time.
- Greater control of cost.
- Dedicated workforce to the one job.

The disadvantages of this approach were seen to be:

- Higher risk of cost variance.
- More effort in monitoring and supervising.
- Limited availability of local concreters to work under a direct-hire agreement.
- Would have to supply all materials including formwork, which the contractor wouldn't fully utilise when construction was complete.

The concrete work was done under subcontract with all materials supplied by the contractor except formwork.

Pipework

The stormwater drainage pipe installation was procured under a hybrid arrangement. That is, the pipe was laid at contract rates, with excavation and backfill being done on a direct-hire basis. This was undertaken this way for the following reasons:

- Detailed/technical work was done by specialists.
- Non technical work was done by direct-hire.
- Greater control of costs and time.
- Greater control of quality; pipework was more closely supervised/inspected and the contractor could be sure backfill was compacted correctly, due to close involvement/supervision of the work.

Earthmoving

It was the contractor's company policy generally not to own earthmoving plant, because of the belief that the cost would be greater to own/lease than to hire the machinery:

- Down time; machinery wouldn't be used continuously.
- High maintenance cost; the contractor would have to set up a workshop with mechanics, fitters etc.
- A large number of plant hirers already existed in the market, and so the market was competitive and the plant would also have a lesser chance of being hired when not being used.
- Good plant operators were hard to find and hard to keep. The machine is only as good as the operator. If the contractor hired plant from a plant hirer it could choose operators and draw on the pool of resources that were already assembled. Some good owner-operators develop symbiotic relationships with contractors.
- Greater flexibility; the contractor could stand-down plant anytime without incurring any more costs. The contractor only paid for the time used.
- It was easier to keep track of direct job costs as the cost stopped when the machines stopped.

EXERCISE

1. What alternative procurement approaches could be considered for the concrete work, pipework and earthmoving? Compare the advantages of these approaches with the one proposed.
2. In what way does the tender amount (cost), contract period (time), and specification and drawings (quality) influence doing the work by direct-hire (and in-house supervision) or by contract?
3. At what stage of the contractor's project should the contractor be considering using direct-hire (and in-house supervision) or contract?
4. Pipework – what are the disadvantages of the approach adopted?
5. Earthmoving – what are the disadvantages of the approach adopted?

2.5.3 CASE STUDY – OUTSOURCING COAL MINING

A coal owner decided to form its own project management team to supervise several major parcels for the construction of a 'greenfield' coal mine. It was intended that the project management team (or at least many key players) would stay on after practical completion to run the producing mine. The scope of work included:

- Provision of infrastructure for a 3 million tonnes per year longwall coal mine (including 4000 m of pit bottom development).
- Procurement of mining equipment.
- Substantial earthworks to create the underground entry.
- Sinking of a 100 m deep ventilation shaft.
- Coal handling facilities to accommodate 900 tonnes/h throughput.

It was decided by the coal owners to outsource the underground works (4000 m of pit bottom development).

Corporate need

Outsourcing permitted time in which the project management team could recruit its own workforce and procure its mining equipment. In particular, the decision to outsource the underground development via contract satisfied a corporate need to buy time for the:

- Formation of its own production mining team (for after the mine was built).
- Recruitment of the production workforce (during the mine's construction).
- Procurement of suitable mining equipment (much would be learnt during construction from the experience and performance of the contractor's equipment).

At the time there were extensive opportunities for experienced miners, and it was difficult for new mines to recruit personnel.

Industrial relations

The coal company had a recent history of conflict with the mining unions. Labour relations had to be improved on this mine. Union coverage agreements were included in the contract, and the contractor took responsibility for industrial relations. This gave a 'clean industrial interface' for the owner.

The issue of contract mining as opposed to direct labour had to be approached in a sensitive manner, because of previous demarcation disputes in the industry.

Because of all the initial groundwork with the unions, the project proceeded smoothly.

Partnering

A formal partnering program was included in the mining contract for the major parcels of work. This had varying degrees of success/ failure. What was being attempted was a pro-active commonsense attitude. The owner believed that if either party decided to take a hard line approach, then all the fancy words would be worthless. The desirable relationship was one that completed the job with minimal conflict, and to the mutual satisfaction of both parties, without an adversarial relationship based upon mistrust and selfish interest.

For this contract, partnering was considered a success. Partnering played a significant role in achieving the resolution of a major dispute. The parties agreed, as desirable, to an out-of-court settlement. Without the formal partnering program, what was assuming the proportions of a contractual impasse at project's end, may never have been broken.

EXERCISE

This decision to outsource the mining was based on several reasons expanded above. In addition there were the following reasons:
• Environment (natural).
• Degree of scope definition.
• Time and cost considerations.
How might each of these have influenced the decision to outsource?

2.5.4 CASE STUDY – LOCAL GOVERNMENT PRACTICES

The dilemma – to use in-house resources or contract-out

Advances in technology, funding reductions, changing needs from construction to maintenance, the economic climate, productivity improvement requirements, the change in community values etc. have brought about a total redirection in the method of the supply of services to the community by one rural council. Council's actions desirably should be based on the adoption of a process that achieves the optimum outcome. The argument used in the past, to have a large in-house workforce, was on the basis of maintaining employment. Although there is no question that unemployment is a major social concern, a business such as a council cannot operate at low levels of efficiency, because the community wants more services at a higher level with less money. Consequently, council is undergoing radical change and, as a result, many of the functions undertaken by council traditionally are being carried out by contractors and consultants.

Traditionally, plant owned by council, and the workforce have been viewed as assets. The council now has taken the approach, that unless certain productivity levels are achieved at the right price, the asset concept is a myth. This does not mean that the individual workers are incompetent or lazy. Unfortunately, councils have resisted technology change by maintaining labour-intensive work methods, rather than using current technology. For example, the tendency has been to maintain the labourer to cut the grass around guideposts with a brush hook rather than using a person with a machine or, further still, using the minimum cost solution of herbicides.

The council now has taken the following approach:
• Work can be undertaken either in-house or by contracting-out, whichever is the better.

An employee or item of plant will be retained by council if defined productivity levels are met.
- The processes adopted must provide target results in respect of cost, quality etc.
- In-house methods and contracting-out have no weighting favouring either option.
- Only a core of in-house staff will be kept. If additional staff are required, the additional staff will be employed for the duration of that need only.

The choice of having an equal weighting for deciding between in-house or contracting-out is deliberate. From experience, both methods can provide successful outcomes. If the reason that the in-house workforce exists is to provide employment only, then it will possibly be inefficient. Similarly, if the in-house workforce is sheltered from technology and is not subject to the influences of competition in a real way, in-house work will be inefficient. Conversely, contracting-out does not always provide cheaper or better solutions. The council's approach is to be in a position to undertake the work by either in-house, contract or a mixture of in-house and contract.

If any of these approaches fail to produce the required outcome, council will then shift to the approach that provides the more favourable outcome. However, to maintain honesty in performance, a combination of in-house work and contracting is in place at all times to undertake the various components of work required in a council operation. No preference is given to any method in terms of social values.

Integral with the approach is planning. To evaluate whether in-house is the more suitable method, the approach begins at the 'ground floor'. This effectively requires a brainstorming approach of the alternatives to achieve the optimum outcome. That is, the job is not matched to the existing labour and plant, but the project is evaluated to obtain the best process and then the required plant and labour for the process are selected.

This approach has required council to:
- Downsize to a core level of in-house plant and labour that can be economically justified.
- Have available on an 'as required basis' various contractors.
- Assess the work that is better carried out by contract.

Despite the reservations of various people, especially those that had the view that a council job was for life, the approach has provided significantly reduced costs. The council's experience has been that significant improvements in in-house performance are achievable, especially when it is perceived that jobs may be under threat at the next review, if the required outcomes are not achieved. Under these arrangements, the majority of council works staff have taken on the challenge, and through a combination of the adoption of modified methods, and use of the most appropriate plant and equipment (council fleet, contractors, hire firms etc.), the council has proven that an in-house workforce can provide an output which is competitive with a contractor. However, in other instances, the council has found that due to the specialised nature or duration of the project etc, or where the in-house resources cannot carry out the work to the standard or price, the use of contractors has provided the more competitive result.

EXERCISE
1. A comparison of productivity by in-house and contracting-out shows that there are cases where contracting does result in a better outcome, and there are cases where it does not result in a better outcome. In the council's case, appropriate productivity has been achieved by utilising con-

tractors, and in-house resources to carry out various functions, thereby creating a situation, by default, of competitiveness. To an extent, a contractor is just another organisation with an in-house workforce. The council considers that an in-house workforce will become more and more difficult to justify, as technology and competition advances further. What is your reaction to these views?

2. Contracting-out has significantly reduced the cost of council's projects. The council's experience is that cost savings up to about 60% can be achieved in certain circumstances, for example in earthworks and rock excavation. This reduction represents a large saving in the construction costs of projects. What is your reaction to such reported large cost savings?

3. By using current technology and engaging construction methods based on efficient processes, the council's experience is that, in certain circumstances, costs can be significantly reduced. For example, the council recently has extensively used cement stabilisation processes for pavement construction. A contractor was engaged to place the pavement material (including cement and water). Council's role has been to compact and trim the finished pavement. The process has reduced pavement costs by about 60%. It was cheaper to undertake the works by a mixture of in-house resources and contractors, rather than carrying out the works totally as a contract. This is because the small production improvements from the contractor are not offset by the additional establishment costs. What is your reaction to this approach?

4. The use of in-house resources for undertaking maintenance grading of gravel roads and shoulders of bitumen roads has proven to be cheaper than contracting-out. The actual hourly cost of ownership, maintenance and operation (excluding operator) of the council's grader is approximately three-fifths that of the most competitive grader available to undertake the work by contract. Since the council's grader is operating approximately 1200 hours annually, and there is no economic advantage contracting the works out after consideration of production and hire rates, the use of in-house resources provides the better solution. What is your reaction to these figures?

5. An open mind is considered to be necessary. Both in-house methods and contracting-out can supply the better cost. However, the council considers that without the competition provided by contracting-out, the reliance on protected in-house resources to provide an efficient outcome is unlikely. Not withstanding this fact, the council considers that one of the major reasons productivity is low in many other council's operations relates to over-specifying, and the failure to accept new and alternative technologies. Therefore many of the advantages that can be achieved by contracting-out cannot be realised. For instance, over-specifying testing requirements can have the effect of slowing down a project to such an extent that the efficiency gains in production are lost in stand-down time. By contrast, the council's aim is to utilise the best method or process, and to utilise the combination of in-house resources and contractors that result in the best outcome. What is your reaction to these views?

6. The council does not consider that there is any particular area, whether it be maintenance or construction, that is not suitable for contracting-out. For example, the council contracts out roadside slashing. This is normally considered an area that is better done by in-house resources. Also, since maintenance is carried out using preventative maintenance techniques, a significant proportion (in excess of 50%) of the council's maintenance is carried out by contracting-out. What is your reaction to these views?

7. One view on the differences between the private and public sectors is as follows. Traditionally, the public sector has had a large workforce to cover rostered days off, sick leave, annual leave etc. It has been slow to accept technology and the use of contractors except for specialised work. However, the private sector generally buys the best technology because it is cheaper than labour, and then contracts out because this has less risk than labour. The private sector then takes on high quality, well-paid core staff to run the technology and the contractors. In the council's case, through the introduction of competition, council is moving closer to the private

sector model. The major difference is that the core staff requirements for a council operation are slightly higher than for the private sector, but through the effects of competition, it is aimed to reduce the differences to a negligible level. What is your reaction to these views?

8. Another view relates to staffing levels between the private sector and the public sector. The private sector has core staff and takes on contractors and casuals for the peaks, rather than providing permanent staff to cater for the peaks. Traditionally, the council's staff have catered for the peaks and, in many cases, in excess of the peaks. In a recent rationalisation program, council reduced staff levels to a core level more approaching a private sector operation, rather than to the core levels applicable to the public sector. What is your reaction to these views?

9. Another view relates to the correct approach and psychology. In the preparation of a tender, the private sector synthesises its bid from the ground up, whereas the public sector analyses from their current position down. This is one of the basic problems many councils fail to consider. Councils tend to benchmark against other councils, who may be less efficient. However, by synthesising the bid and the adoption of best practice, benchmarking is not as important, because the contractor has benchmarked against performance able to be achieved. The council adopted this approach a couple of years ago for the preparation of bids and also to restructure the workforce. The adoption of this approach provides a rational basis for rightsizing the in-house workforce, and provides production targets based on detailed technical calculations for use by the workforce. Utilising costing data based on past performance, as traditionally has been the case in the public sector, is in the council's belief a 'breeding ground' for inefficiency. What is your reaction to these views?

10. Although recognising that there are cost savings to be made, many other councils are resisting change. Their argument appears to be based purely on the social issue of employment. Although it is agreed that unemployment is a social concern, the council does not consider that operating a business such as a council inefficiently helps the problem significantly. In fact, from experience, if a council operates efficiently, and since the budget does not generally reduce as a result of gaining efficiency, the total number of jobs including in-house and contracting is often more than if the council is operating inefficiently. The major difference is that the jobs are not all in-house. However, the amount of work undertaken through efficiency improvements is significantly greater and therefore the gains significantly outweigh the losses. What is your reaction to these views?

2.6 EXERCISES

Exercise 1

a) Why contract? Why not contract? Under what conditions, other than mentioned above, might an owner use in-house procurement?

b) Why has routine maintenance been traditionally done using in-house resources of an organisation?

Exercise 2

For public sector bodies, community service obligations were always paramount. Does the notion of becoming 'efficient' conflict with these obligations? Can organisations provide traditional services, while becoming economically efficient? Why are not traditional services costed?

Exercise 3

The following issues were raised in an enquiry on the subject of contracting-out by government agencies (IEAust, 1995). Give your responses.

a) What impact does contracting-out have on government control over the provision of services?
b) Does contracting-out enhance or detract from the skills base available to government?
c) What impact does contracting-out have on the quality of services provided?
d) What effect will contracting out infrastructure, such as water services and telecommunications, to international consortia have on the opportunities for [local] engineers and technologists to keep up with world best practice?

Exercise 4

AFCC, a contractor's organisation, argues strongly against in-house work (AFCC, 1972). Additional to the points raised above against in-house work, it is argued (albeit without proof):

- Doing work in-house leads to ... *poor utilisation ratios for specialised plant and equipment ...*
- *[In-house] methods waste human resources through lower productivity.*
- *The contract [approach] is private enterprise at its best – virtually pure competition; entrepreneurship in action; cost savings that are passed on to the [owner]; the cheapest way of effecting the work; controlled quality and performance, governed by the specifications, supervision and the need of the contractor to maintain a good reputation; and the maximum incentive for faithful and timely completion.*

Give your views on these three points raised.

Exercise 5

Household garbage collection is now commonly done under contract, with contractors paid on a piece rate, that is payment is per garbage bin emptied. Formerly, local government employed its own labour to do the garbage collection. Trucks with mechanical arms empty the contents of the garbage bins into trucks, operated by a sole driver/operator. There is no manual handling.

Payment is based on bins emptied, irrespective of how much spillage occurs on the roadside. Hence contractors rush around the streets emptying as many bins as possible in as short a time as possible. As expected, the streets end up with rubbish scattered all over them through spillage or not emptying the bins correctly into the trucks.

How could you better formulate the contract, in order that the bins are emptied economically, yet the drivers/operators take care not to spill the rubbish?

Exercise 6

How do you address probity issues when an in-house group is tendering for work along with outside contractors?

Exercise 7

a) Compulsory Competitive Tendering (CCT) legislation requires that all Victorian (Australia) councils market test a set percentage of their services against private and public sector competition. This has spurred many councils on to not only retain existing work, but also win contracts from other municipalities. Staff were trained in business management, planning, costing, budgeting and other skills needed to compete successfully in the tendering process. Issues of plant rationalisation, resources, marketing, promotion, finance, work practices and performance indicators were looked at. Some municipalities joined together to give efficiencies of scale.

Do you see this as desirable that councils should go this way? What flow-on effects could you expect for the complete operations of councils? Should councils not go this way, but rather contract-out anything that the private sector could do better? Give your views.

b) CCT followed from the National Competition Policy. New South Wales politicians, in contrast to Victoria, viewed CCT as a 'heavy-handed' reform. They saw competition policy, '... not as some mandatory set of guidelines, but as a further incentive to undertake reform such as selective tendering and establishing a more even playing field for competitive evaluation of in-house provision of services versus contracting-out.' (Local Government Focus, July 1996, p. 8) What are your views?

c) In promoting greater efficiencies in the public sector, the New South Wales Premier's Department gives the following examples of how better value for money was obtained for the community by contracting-out (Office of Public Management, 1991):
*   *The Water Board obtained over 100% productivity improvement in meter reading.*
*   *The Commercial Services Group obtained a 60% improvement in labour productivity in the Q Stores [a materials supplier].*
*   *The Northern Area Health Service saved $2 million per annum in hospital cleaning.*
*   *Overseas, the United Kingdom civil service has made cost savings on average of 25%.*

Do you expect that internal efficiencies could have been made to match the above quoted improvements obtained by contracting-out? Or would you expect that, because of restrictive work practices and related constraints, the only way to achieve significant efficiencies would be by a complete break and go from in-house to contract?

Exercise 8
List classes of activities which you consider would be candidates for contracting-out.

REFERENCES

Antill, J.M. 1970. *Civil Engineering Management*, Angus and Robertson.

Australian Federation of Construction Contractors (AFCC) 1972. *The Case Against Day Labour Construction*.

Butler, K. 1995. Lessons in Partnering, *CII Annual Conference*, Melbourne.

Fraser, L. 1992. Competitive Tendering and Contracting-Out of Local Government Services in Australia, Public Sector Research Centre, The University of New South Wales, *PSRC Discussion Paper* No 26, Sydney.

Institution of Engineers, Australia 1995. Re-Engineering Competitiveness Inquiry, *Issues Paper*.

NPWC/NBCC 1990. *No Dispute,* National Public Works Conference, National Building and Construction Council, Canberra.

Office of Public Management, New South Wales Premier's Department 1991. *Competitive Tendering and Contracting Out – Guidelines*, 23 pp, Sydney.

VicRoads 1994. *Australian Project Manager*, 14(3): 23.

BIBLIOGRAPHY

McInnes, N. 1997. *Bidding to Win – Complete Approach*, Public Works Engineering.

CCF 1996. National Conference Report, Competition in Contracting: debating the issues, *The Earthmover and Civil Contractor*.

Testing time for road maintenance contracts 1995. *The Earthmover and Civil Contractor*.

Giummarra, G. 1997. Competition through innovation, *Highway Engineering in Australia*.

CHAPTER 3

Contract (payment) types

3.1 INTRODUCTION

Different contract (payment) types offer different incentives to the contractor or consultant and influence the cost, delivery time and performance on contracts. 'Payment' here refers to the overall way the contractor or consultant is reimbursed or paid for its role in the contract.

On any given project with many contracts (between the various players or stakeholders – owner, consultants, contractors, subcontractors, suppliers, ...), there may be a whole range of contract (payment) types, each chosen to reflect the circumstances particular to the nature of the work.

This chapter explores the different contract (payment) types and examines the conditions under which each might be most suitable.

Terminology

A disincentive may be thought of as a penalty. However, the law may not allow contract wording involving penalties, and so to achieve the desired effect or result of a penalty, the contract wording may have to be carefully phrased such that a penalty becomes, perhaps, an unattained bonus or similar, that is, a disincentive.

The disincentives/penalties referred to below reflect the result. Care of their wording in contracts may be necessary.

3.2 OUTLINE

Contracts fall into two groups (Fig. 3.1) classified according to the form the consideration from the owner takes. 'Consideration' is something of value given in return for something else. Here consideration refers to the payment from the owner to the contractor.
- Where the consideration is either a stipulated sum of money covering all the work or is a set of monetary rates covering the components of the work, the contracts are referred to as *fixed price contracts*.
- Where the contractor is paid the cost of the work together with an additional amount for the use of the contractor's services, the contracts are referred to as *prime cost contracts*.

As well, it is possible and common to have contracts which are a combination of fixed

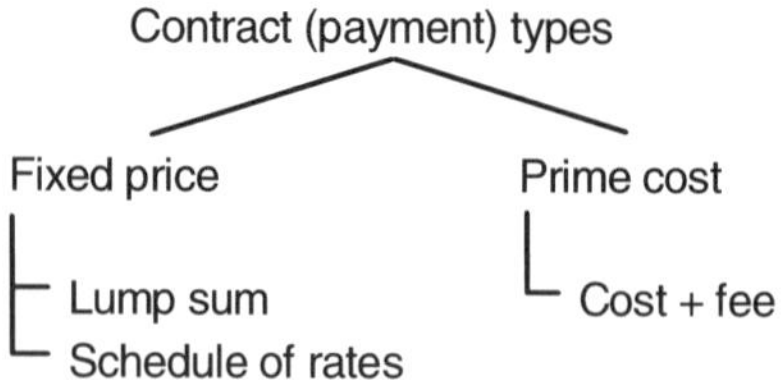

Contract (payment) types

Fixed price   Prime cost

├─ Lump sum   └─ Cost + fee
└─ Schedule of rates

Figure 3.1. Contract (payment) types.

price and prime cost components. Popular usage might describe the contract after the predominant form of payment type used in the contract, although different payment types might be present in the one contract.

Convertible contracts begin on the basis of one payment type and convert to another payment type at a defined point in the project. The second contract may be subject to competitive bidding, or the one contractor may carry out all the work, if a sufficiently good relationship exists between the owner and the contractor. For example, the project may start on a cost-plus basis, and convert to a lump sum once the scope can be sufficiently well defined.

Usage

There are a range of situation-dependent factors that determine which contract (payment) type or combination of contract (payment) types is most suitable in any given case. It is a case of 'horses for courses'.

... we believe that ... fixed price should be the norm, and it should rest with those advising the [owner] to prove that some other form of payment ... would be in [its] interests. (The Aqua Group, 1975) However, this is only one view.

A general belief is that the owner should, wherever possible, define the work as carefully as possible beforehand, irrespective of the type of contract used. This could be expected to reduce the risks to all parties.

Transfer of financial risk

There is an increasing transfer of financial risk from the contractor to the owner on going from the top to the bottom of the following boxed list. As more risk is taken by the owner, it could be expected that the contract price would decrease. There may be an overall saving to the owner by accepting some risk. Generally there is an expectation from owners that contractors should bear some risk, though some owners take this to an extreme and (through choice of conditions of contract, including payment type, and delivery method) ask contractors to bear essentially all risk.

Fixed price
• Lump sum
•   no escalation/rise and fall; no delays except owner-caused
•   no delays except owner-caused

- • no escalation/rise and fall
- • with escalation/rise and fall; with delays
- • with escalation/rise and fall; with compensation for delays
- • Schedule of rates
Prime cost
- • Cost-plus
- • fixed fee
- • percentage fee

Public accountability may inhibit public sector bodies from using anything other than fixed price contracts. Owners with limited financial resources may not wish to take the risk associated with cost reimbursement contracts, even though they may be more suited to the type of project.

Of course, there are risks other than financial risk that the owner has to consider.

From the contractor's viewpoint, there needs to be thought given to an owner's financial resources and the matching of the payment type to these resources, and thought given to the contractor's financial resources and the risks it is prepared to take. Contractors with limited resources may be more inclined to take work with a reasonably certain return.

3.3 FIXED PRICE CONTRACTS

Fixed price contracts are common in project-based industries. Many organisations, industry interest groups and standards bodies publish general conditions of contract intended for fixed price contracts. Fixed price contracts promote competition between tenderers, and hence are favoured by owners, be they from the public or private sectors.

The traditional delivery method commonly uses fixed price contracts.

Adjustments

In tendering for a fixed price contract, the contractor accepts the risks for the stipulated sum or stipulated rates nominated in its bid. However, there may be provision for adjustments:

- • Costs of labour and material that change or fluctuate (called *rise and fall* or *escalation*) over the duration of the contract; this is commonly based on an agreed formula related to a published index.
- • *Variations* (or *extras*), changes or adjustments to the scope of the work, commonly at the instigation of the owner, or because of something unforeseen; owners, however, don't have an unlimited power to order variations – they must be reasonable. Some contracts make no provision for variations and their payment, in which case, it is not reasonable for the contractor to perform variations for which no payment will be received.
- • *Compensation* to the contractor *for delays* caused by the owner. Delays caused by the contractor, or delays due a neutral source could be expected to receive no compensation from the owner.
- • *Provisional items*, known to be part of the contract but for which complete details are

not known at the time of tendering; the contractor allows a (possibly nominated) *provisional sum* to cover these items; typical examples include earthworks and mechanical services; the contract price is adjusted up or down and the contractor is reimbursed according to the actual cost of the item relative to the original provisional sum; allowance may be made for profit and attendance.

* *P.C. (prime cost, pc) items*, mentioned in the contract documentation, but yet to be selected by the owner; the contractor allows a (possibly nominated) sum (*prime cost sum*) in its tender for these items; typical examples include bathroom fittings; the contract price is adjusted up or down and the contractor reimbursed according to the actual cost of the item relative to the original prime cost sum; no allowance may be made for profit. Prime cost items may also refer to packages of work to be carried out, for example by nominated subcontractors. [Some people use the terms 'provisional items/sums' and 'prime cost items/sums' interchangeably.]
* *Latent site conditions*, conditions not anticipated at the time of tendering, e.g. unexpected ground conditions.
* *Attendance* of a contractor to nominated subcontractors; the service provided by a head contractor.

Such provisions may mean that the final cost of the work is different to the tendered price. But generally the owner knows to reasonable closeness what the work is going to cost and can budget accordingly. The total contract price is accordingly 'almost fixed', rather than being 'rigidly fixed'.

Terminology

Note, some writers use the expression 'fixed price contract' or 'firm price contract' in a rigid sense, that is there are no provisions for changing the original contract bid. The term 'firm-fixed-price' (FFP) is also used.

The 'fixed price' definition adopted in this book follows Antill (1970): ... *they are rightly termed fixed price, because once the offer is accepted, the contractor cannot change [its] price or rate for the work.*

Some people use the terms 'firm price' and 'lump sum' in the sense that there will be no adjustment for variations, remeasurement, or fluctuations in the cost of labour and materials. This usage is not followed here. However care needs to be exercised when talking to people or reading documents as to the intended definition of the terms used.

Some other apparent anomalies are as follows. Provisional sums and P.C. sums are cost reimbursement in nature, yet they sit within a fixed price contract. Nominated and other subcontracts (supply or work), while they may be fixed price between contractor and subcontractor, may be on a cost-reimbursement basis between owner and contractor. In a contract for cost reimbursement of labour, materials, work, ... (including day labour or daywork), the contractor's head office overheads and profit are fixed price, even though the work may vary from that initially envisaged; the contractor makes an estimate of anticipated overheads and expected profit before work commences. To summarise, fixed price contracts could be expected to have cost-reimbursement components, and cost-reimbursement contracts could be expected to have fixed price components.

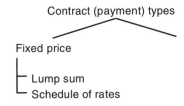

Figure 3.2. Categorisation of fixed price contracts.

Classification

Fixed price contracts are classified as:
- *Lump sum contracts* (also called stipulated sum contracts, and contract in gross),
- *Schedule of rates contracts* (also called unit price contracts, measurement contracts, item rate contracts, bill of quantities contracts, piece rates, schedule of prices, and measure and value),

depending on the method used to price the work (Fig. 3.2).

(Note: some people use the terms 'lump sum' and 'fixed price' synonymously, to mean what is defined as lump sum here.) A schedule of rates contract is legitimately called a fixed price contract *because the basis of payment has been predetermined – the price being fixed, only the quantity of work is unknown and this is to be ascertained by measurement as the work is done. This is still true even if the rate varies with the quantity done or used* (The Aqua Group, 1975).

Lump sums may apply not only to complete contracts, but also to sections or packages of the work.

It is possible to have contracts which are a combination of lump sum and schedule of rates components. The known defined work is covered by the lump sum component; the known less well-defined work is covered by the schedule of rates component. For example, the construction of a building or bridge may be done on a lump sum basis, while the foundations may be done on a schedule of rates basis; on a hydroelectric project, the power house structure may be built on a lump sum basis while foundations, dam structure and tunnel work may be done on a schedule of rates basis.

Clear delineation needs to be made in the contract as to what work is covered by the lump sum, and what work is covered by the schedule of rates.

Terminology

Schedule of rates

A list of unit items of work priced at a rate per unit. The schedule is used in conjunction with the measurement of work to calculate payment, under a schedule of rates contract. Sometimes called a *schedule of quantities and rates*, or a *price schedule*.

Tenderers' rates in a schedule of rates contract are usually based on approximate quantities.

It is not uncommon to see a schedule of rates in a lump sum contract to provide for the payment of work (variations, extras) additional to that included in the contract. This minimises disputes over payment valuation of the additional work. Lump sum contracts that allow for variations according to a schedule of rates, may be incorrectly referred to, by some people, as schedule of rates contracts.

Bill of quantities

The measured/quantified description of work in a contract.

For lump sum contracts, the bill of quantities (together with included rates) can be used for progress payments and for valuing variations or change orders. Such a bill of quantities may be 'marked up' by the contractor, with rates, only after the successful contractor has been selected.

A bill of quantities contains reasonably accurate quantification whereas, in a schedule of rates contract, accurate quantification takes place at a later time to determine the payment to the contractor.

However this distinction is muddied by people referring to an approximate bill of quantities or to a schedule of approximate (provisional or partial) quantities.

Note also that some people use the terms 'schedule of rates' and 'bill of quantities' interchangeably, while others consider them as separate documents. In such circumstances, it is better to look at the intent of the document or the intent of the usage of the term, rather than its naming.

3.3.1 LUMP SUM CONTRACTS

Where the scope of the work can be well defined, the documentation is reasonably complete and few alterations or changes to the work are anticipated, the lump sum form of contract may be favoured by an owner. The contractor accepts the risks inherent in, or as defined in, the tender documents and offers to do the work for a stipulated sum of money. This stipulated sum would be expected to include all project direct and indirect costs, contingencies, head office overheads and profit/margin/markup.

The contractor stands or falls based on its performance relative to the lump sum price that it bid; the risk associated with completing the work at the tendered price, in the tendered timescale and to the required standards lies with the contractor. The contractor's tender should accordingly reflect this risk, though in a competitive tendering environment, contractors may (unwisely?) elect not to price in some risks for fear of being non-competitive. Tenderers accept the situation that there may be more involved in the work than they have allowed for in their tender, although a lot of such unknowns can be removed through the inclusion of special contract clauses, for example covering 'rise and fall', or guaranteeing quantities.

Where the work contains parts which cannot be fully detailed at the tender stage, provisional sums may be included. Alternatively, a schedule of rates contract, or prime cost contract may be a better option. A lump sum contract would not be the preferred type for situations where the scope cannot be completely delineated beforehand. A schedule of rates contract may have some large single items, which effectively are lump sum components for these portions of the work.

The owner can be reasonably confident of the approximate total cost, although provision is usually present to adjust the cost for changes in scope etc. Where the tendering process is competitive, the owner can feel confident of obtaining a reasonable price. Pub-

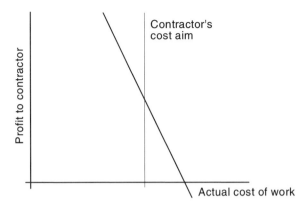

Figure 3.3. Lump sum contract.

lic sector work is commonly carried out using lump sum arrangements because of the better certainty it gives in terms of completing the project within budget.

A fully defined scope of work minimises variations and delays and consequential increased costs. The contractor knows exactly the work required and can plan the operations and budget to suit. There is an incentive for the contractor to operate efficiently, in order to increase its profit; conversely, where the contractor operates inefficiently, there is the potential to make a loss (Fig. 3.3).

Lump sum contracts may be used for any type of activity, or combination of activities, including investigation, design, supply, execution (construction, fabrication, manufacture), management, maintenance etc. The difficulty however may be in defining the scope of work accurately enough. The owner has to specify exactly its goals and requirements, for a lump sum contract to work best.

Contractors are usually expected to carry the cost burden of preparing their tenders. The work involved in developing a proposal to a level where a competitive lump sum price may be tendered can be substantial, and can involve a substantial cost. This preparation cost is an overhead that gets incorporated into the bid. Owners may reimburse contractors on some projects. Given that contractors may only win, a small fraction of the tenders that they submit, many would argue that the cost of tendering should be borne by the owner as it is the owner who is getting the benefit from competitive tendering. The rational financial argument would be that costs generated by a project should be borne by that project and not spread over other projects.

Where site investigation is involved, this might be carried out by the owner, and would be recommended for the owner to carry out, in order to make the tender documents as complete as possible. Tenderers could be expected to be reluctant to invest in site investigations where there is only a small probability of their winning the tender. One reason why owners might be reluctant to guarantee site information is for fear of future claims, should the information turn out not to be accurate. However, not guaranteeing the site information may lead to the tenderer loading its price to cater for uncertainties.

It is not recommended that a lump sum contract be used where the lump sum cannot be enforced, for example where the true extent of work is unknown at the start. The owner becomes deluded as to the project cost. As additional work becomes known, the contractor has a case for extra reimbursement. If the work turns out to be less than originally allowed for, the owner generally does not benefit from this. There is also the potential for, and cost of, engaging a new contractor, should the first contractor perceive a loss and walk away from the project.

Progress payments to the contractor only need to be approximate, because the total sum paid to the contractor will not exceed the bid amount.

Bill of quantities

A bill of quantities may be provided by the owner:
- For insertion of unit pricing or rates, to compare alternative tenders, and for valuing variations.
- For the convenience of the tenderer, listing every work item.

However, the bill of quantities may be stated to not form part of the contract, and its accuracy not guaranteed, but is provided as a guide for tenderer's information purposes only. Tenderers are then responsible for making their own work quantity estimates. Where the bill is considered to be part of the contract, the owner takes the risk associated with errors, omissions, discrepancies etc in the bill, and the consistency of the bill with other contract documents. For this reason, some owners let the tenderers develop their own bills, thereby transferring the associated risk.

To avoid the trouble of producing a bill of quantities, some contracts require only to measure quantities which vary from the intentions as shown on the drawings.

An alternative, for valuing variations and valuing progress payments, is for the tenderer to provide unit prices or rates.

Summary – lump sum contracts

Owner's position – advantages

- A good indication of the cost of the work may be obtained before work commences; a reasonably assured total price.
- Minimal involvement in contract administration; detailed accounting and measurement avoided – total payment fixed, progress payments based on estimated percent complete not measurement of actual work done; minimal monitoring.
- Price competition possible.
- Most of the risks associated with the work are carried by the contractor.
- There is an incentive for the contractor to work efficiently.

Owner's position – disadvantages

- The documentation must be fully complete at time of tendering; lead time is the longest of all contract (payment) types.
- Owner and contractor placed in adversarial positions; potential for disputes is high – opposing interests in the final cost.
- Reduced flexibility to make changes; alterations and additions (change orders) and unforeseen problems can cause trouble.
- Changes in work or unforeseen difficulties may end in disputes, with an associated ex-

tra cost; close administration required to prevent possible overpricing of variations; heightens adversarial relationship between owner and contractor; extra costs possible even though a fixed price contract.
- Owner has little influence over the contractor's approach.
- If low profit levels, materials and workmanship may suffer; maintaining quality requires good documentation and inspection; accepting lowest bid may result in the contractor using marginal subcontractors; prequalification of tenderers can address this partly.
- Ambiguities, discrepancies and omissions in the design documents can be exploited by an unscrupulous contractor.
- Close administration of the extension of time provision required so as not to forfeit the right to claim liquidated damages for the contractor's delay in completion.
- If the contractor fails to perform, the remedies available have to be administered delicately.

Contractor's position – advantages

- Innovative or efficient contractor can improve its profit.
- Contractor's bid reflects expected profit level; contractor can name its own price.
- Minimum involvement of the owner in the details.

Contractor's position – disadvantages

- The cost and time in preparing a tender can go unrewarded.
- Potential for disputes is high; owner controls the money related to disputes.
- Changes in work or unforeseen difficulties may end in disputes, with an associated extra cost and possibly delayed payments in recovering costs.
- Most of the risks associated with the work (e.g. related to weather, strikes, external factors) are carried by the contractor, some may not be under the contractor's control.
- Fair/equitable price adjustment formula difficult to establish.
- Marginal subcontractors may have to be used in order to be price competitive.

3.3.2 SCHEDULE OF RATES CONTRACTS

Where the nature and scope of the work are known but quantities are indefinite, the owner may prefer to use a schedule of rates contract. In a schedule of rates contract, the contractor includes unit rates in its bid for each work item. As the work is completed, the quantities are measured and the contractor gets paid accordingly.

Examples
Subsurface and earthworks, for example excavations and foundations, may commonly be done on a schedule of rates basis because of the uncertainties involved in below-grade work. Heavy construction work is commonly done on a schedule of rates basis, because quantities can be difficult to establish in advance of the work. Redecoration work is also suitable. Schedule of rates contracts are also used, for example, in open-cut mining, underground mining, diamond drilling, rehabilitation, tailings dam construction, and explosives supply.

> Schedule of rates contracts are particularly well suited to heavy construction works, such as highways, bridges, dams and river and harbour improvements, where large quantities of a relatively few types of construction processes are involved.

Some organisations use schedule of rates contracts almost exclusively. They argue, for example, that changes in foundation work for building structures are common, while in major work such as motorways, the final drawings and specifications commence more than a year before inviting tenders, and changes become inevitable. To wait for the precise definition needed for lump sum contracts is considered neither practical nor necessary; the work would be delayed and extra documentation costs would be incurred, with no real guarantee in increased certainty regarding the final project cost. However, this thinking should not be used to excuse poorly drafted tender documents, including the drawings.

Typically the contractor tenders against a schedule which contains estimated or guide quantities against each item, and a standard method of measurement would be used to describe the quantities. Tenderer selection is based on the estimated quantities multiplied by the tendered rates, and summed over all work items (see Fig. 3.4). Work items not expressible in quantities may be called 'jobs', 'items' or similar, and single prices (lump sums) are tendered against these.

A different approach involves the owner developing a pre-priced schedule, with the tenderers adjusting these prices up or down to reflect the tenderer's costs to do the work and the tenderer's bidding strategy. The intent of this approach is to make the tender evaluation a bit more objective.

Care has to be exercised that no work items have been left out of the schedule by the schedule drafter. Provisions may exist in the contract conditions for the negotiation of new rates for unforeseen work, or the changing of rates should the final quantities be markedly different to those given in the original schedule. The contract conditions need to be fair in terms of, for example, how significant quantity variations or latent conditions are handled.

To cover against the possibility of different degrees of magnitude of the quantities (guide quantities compared with final quantities), the unit rates may be on a sliding scale (for example, the rates may decrease as the quantity increases) or variable scale (for example, if quantity is $\leq X$, then rate is y; if quantity is $> X$, then rate is z). Alternatively, there may be agreed limits of accuracy of the estimated quantities appearing in the tender documents. Where the actual quantity deviates significantly, for example more than 10% (or even maybe 20%), from the guide quantity, the contract may allow for the rate to be negotiated or determined by the owner; where the actual quantity exceeds 110% of the guide quantity, the rate could be expected to reduce, and where the actual quantity is less than 90% of the guide quantity, the rate could be expected to increase. Larger quantities

Item No.	Quantity	Unit	Description	Rate ($/unit)	Total (quantity × rate)
...					
45	150	m^3	Clay excavation to 1 m	–	–
46	36	m	300 mm dia pipe	–	–
...					
Total					Σ

Figure 3.4. Example part of a schedule.

may allow economies of scale for the contractor; the proportion of one-off indirect costs (e.g. mobilisation and demobilisation) to direct costs could be expected to decrease with increased quantities.

The contractor is bound by the tendered rates, and not by the quantities.

Rates tendered usually are the direct cost of doing the work increased by adding (spreading) the contractor's indirect costs (overheads, profit, ... – both one-off and time varying).

Unbalancing

At the time of tendering, the owner, ideally, has stated the approximate magnitude of the quantities and their order of accuracy. Where the contractor perceives that the *quantities* given in the tendered documents are inaccurate, the contractor might *unbalance* the tendered rates so as to, either increase potential profit, or give a lower total tendered price, or both. For example, on items which the contractor believes the final quantity will be greater than that stated by the owner in the tender documents, the contractor might tender a higher rate than usual. And on items which the contractor believes the final quantity will be less than that stated by the owner in the tender documents, the contractor might tender a lower rate than usual. That is, there is a risk to the owner if its estimates of quantities are wrong or unexpected work arises.

Unbalancing of rates might also be carried out in a *time* sense; for example 'front-end loading' involves the contractor increasing the rates (over what it would normally bid) of items that occur early in a project, and counterbalancing this with decreased rates for items that occur late in a project (for example, 'close out' items), in order to improve the contractor's cash flow from larger progress payments early in a project. In Figure 3.5, front-end loading increases the early progress payments, and moves the stepped income curve closer to the cost curve, thereby requiring the contractor to find less finances to fund

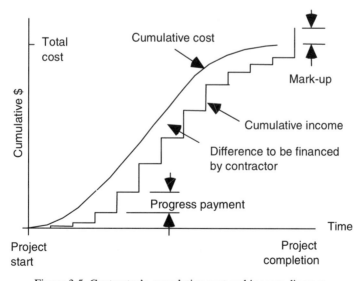

Figure 3.5. Contractor's cumulative cost and income diagram.

the project, and improving the contractor's cash flow (costs minus income). [Of course, there are other ways that a contractor can improve its cash flow, for example by getting an up-front (pre-start, mobilisation, front money, ...) payment from the owner, delaying payment to subcontractors and suppliers etc. And there are projects where the owner finances the work instead of the contractor.]

The downside from the contractor's viewpoint is that the contractor can lose out if the estimated quantities or the project duration are different to that anticipated (by the contractor) at the time of tender. This is because the tendered rates contain the contractor's indirect costs, and these may not be fully recovered in the case of changed quantities or project duration.

Unbalancing in a time sense can also occur in lump sum contracts where the contractor is asked to provide rates as a basis for payment.

General

Many people regard a schedule of rates contract as the fairest of all contracts. The contractor gets paid for all the work done, while the owner only pays for the work done. For example, in driving piles, a contractor tenders a price per meter based on the best available field information provided by the owner, the ground conditions are not known exactly prior to driving, and the contractor is paid for the driven length.

The owner has the convenience and flexibility of changing the scope of work, often without altering the contract documents. Only work so directed by the owner is paid for. Undirected or unapproved work done by the contractor is not paid for.

The contractor's bid does not contain any contingencies such as may be present in a lump sum contract bid, contingencies which are there to cushion the contractor against any uncertainties and unknowns. Possible uncertainties and unknowns might be inaccurate quantities or unknown site conditions.

As with lump sum contracts, the tendering process for schedule of rates contracts can be competitive. Bids can be called earlier, however, and drawings issued as work proceeds. No time is wasted up-front fully documenting the work or working out accurate quantities.

On the negative side, the owner only knows the final cost of the work at the end, and additional administrative staff are needed to measure the quantities as the work is done. Cost control has to be done in terms of predicted costs, which are not as definite as for lump sum work. Regular measurement is necessary in order that valuations are accurate and up to date, and costs are accurately accounted for. For cost control purposes, the work might be divided up into isolated packages, as small as can be conveniently handled; this facilitates the tracking of variations.

Schedule of rates contracts, in conjunction with performance-based specifications, make the evaluation of competing tenders, possibly based on different proposals, difficult. Once the tender is accepted, however, quantities can be developed for cost control purposes.

It is common for contract parts, which are not as well defined as others, to be on a schedule of rates basis, while perhaps the rest of the contract is on a lump sum basis.

Measurement, the method of measurement and remeasurement (assuming the measurement is not accepted) of work done may lead to differences of opinion between the owner and the contractor. All issues, including who pays for measurement and remeasurement, need to be covered in the contract's special conditions.

Case example

A formwork contractor quoted a price per square meter, that is on a schedule of rates basis. (Formwork, in this case was plywood supported by timbers and props. The formwork was for a concrete slab.) On completion of the work, the owner measured the number of square meters of formwork used, multiplied it by the contractor's quoted rate, and accordingly paid the contractor. The contractor queried the magnitude of the payment, and proceeded to measure the area of the slab.

Because the slab was supported on walls below, the area of formwork was less than the area of the slab. A misunderstanding had occurred over the rate quoted by the contractor and accepted by the owner.

Summary – schedule of rates contracts

Owner's position – advantages

- Payment is fair and equitable; payment is for work done; there is less chance of disputation compared to lump sum.
- Small involvement in contract administration.
- Does not necessarily require quantities, but does require scope definition and pricing schedules; lead time is less than for lump sum contracts.
- Fast-tracking possible.
- Price competition possible; tenderers can be compared on the same basis.
- Some flexibility to make changes; freedom to alter the work to suit project conditions without contract variations or change orders.
- There is an incentive for the contractor to organise work methods efficiently.
- Less likelihood, than in lump sum contracts, for the contractor to include contingencies in its bid; possible keener price than a lump sum bid.

Owner's position – disadvantages

- Shared risk with the contractor.
- Final cost and time only known approximately; measured quantities may be greater than estimated.
- Owner has only a small influence over the approach of the contractor.
- Changes in work or unforeseen difficulties may end in disputes, with an associated extra cost.
- Cost and time overruns can occur.
- Quality depends on the level of monitoring and inspection.
- Supervision costs involved in measurement and associated paperwork – progress payments based on measurement of work done; possible disputes over measurements and method of measurement; where there is disagreement over the measurement, who pays for remeasurement – shared or owner's cost? Measurement has to be done exactly because it is the basis of payment to the contractor.

- Not suitable for design-and-construct delivery.
- Quantity unbalancing to the detriment of the owner possible, if its estimates of quantities are wrong or unexpected work arises.
- Front-end loading possible.

Contractor's position – advantages

- Payment is fair and equitable; payment is for work done.
- Previous work pricing can be adopted for similar work.
- Small involvement of the owner.
- Profit assured.
- Quantity take-off need only verify the guide quantities.
- Quantity unbalancing possible.
- Front-end loading possible.

Contractor's position – disadvantages

- Shared risk with the owner; contractor may bear risk of inclement weather, strikes, and other external factors.
- Small cost and time in preparing a tender.
- Changes in work or unforeseen difficulties may end in disputes, with an associated extra cost and possibly delayed payments in recovering costs.

Bill of quantities contracts

Some people make a distinction between 'schedule of rates' contracts and 'bill of quantities' contracts. This distinction is not followed in this book, but the distinction is reported here for reference purposes.

[Note also that some people use the terms 'schedule of rates' and 'bill of quantities' interchangeably, while others consider them as separate documents. In such circumstances, it is better to look at the intent of the document or the intent of the usage of the term, rather than its naming.]

The essential difference, made by these people, is one of accuracy of the quantities given to the tenderers. 'Bill of quantities' contracts have quite accurate quantities (coming from increased dissection of the work); 'schedule of rates' contracts have only broad estimates of quantities. In all other matters the two types of contracts are essentially the same.

An example of the greater dissection of work that may be found in a bill of quantities contract is that of constructing concrete elements. Here quantities may be given for formwork area, concrete volume, steel reinforcement itemised, and surface finish. This assists progress payments but may be costly to set up and measure. A schedule of rates contract, by comparison, may be just in terms of an all-up measure (e.g. cubic meters) for such elements. The measure may be a suitable indicator of the work involved provided, for example, the proportions of the element are not changed by the owner.

The total sum, tendered under a *bill of quantities contract* is the sum of the individual items as priced in the bill, including any prime cost sums, lump sums, and provisional sums. For the purpose of tendering, the quantities placed against the items showing the amount of work to be done are quantities measured from the contract drawings. The quantities are not estimated; they are measured as accurately as possible from the drawings. When the work is constructed, the quantities are replaced by the measurement of the actual quantity of work that the contractor carried out under each item. Then it is an accurate calculation.

Schedule of rates and quantities. People who refer to bill of quantities contracts have the following interpretation of a schedule of rates contract.

There are some operations where it is not possible to put into the bill of quantities, measurements of quantities based on the contract drawings. The full extent of the work to be done cannot be foreseen at tender time. An example is a contract for the sinking of a borehole for a water supply; it is frequently not possible to state in advance how deep the borehole must go in order that it will produce a given quantity of water. There are other occasions when it is needful to start the work of construction/fabrication/erection before the design drawings are ready, i.e. before measurement of quantities can be made from such drawings. In these instances, the contract can be based on a schedule of rates and quantities. The schedule of rates and quantities is similar to a bill of quantities; but it is not the same as follows:

- Quantities against the individual items are either not inserted, or they are entered in estimated amounts or in round-figure provisional quantities.
- More items are scheduled for temporary work than usually appear in a bill of quantities, e.g. items such as setting up plant etc. The amount of temporary work that the contractor will have to undertake is uncertain.
- The remainder of the scheduled items tend to describe operations by the contractor rather than outputs, and the number of items is less than in a bill of quantities.
- There is no implication given that all or any of the work scheduled will in fact be carried out.

It is important that the schedule is clearly headed 'Schedule of rates and quantities' so that the contractor is forewarned that all the quantities and items must be considered as not necessarily final. The tenderer quotes against each item in the unit prices on the possible basis that multiples of the amount of work estimated might be done, or none of the work in the item might be done, and does not incur a loss of money whichever way the actual quantities go.

This is different from a bill of quantities contract, where there is an implication that the total work undertaken will not be so substantially different from what is delineated in the contract that a contractor could maintain it had been misled as to the size of the contract.

In a bill of quantities contract, the indirect costs and profit to the contractor can be spread over the bill items (excepting the provisional items) as the contractor chooses, in the knowledge that the great majority of these items will be carried out.

In a schedule of rates contract, there is no guarantee that all or any given proportion of items will be carried out. Therefore, each item must carry its own overheads and bring the contractor adequate reward if undertaken in large or small quantities, irrespective of the amount of work done under other items.

A schedule of rates contract can be fair to both parties. Such contracts do not give the same assurance in regard to total cost to the owner as a bill of quantities contract.

3.3.3 GUARANTEED MAXIMUM PRICE (GMP) CONTRACTS

The term guaranteed maximum price (GMP) is used to mean different things by different people. A common view is that the price quoted by the contractor is the upper limiting amount that will be paid by the owner (unless the owner instigates changes). There is no potential in the contract for extra payments due to such matters as incomplete documentation, nominated subcontractors, latent conditions, industrial relations, extensions of time, delay costs and variations. In effect, all risk is being transferred to the contractor, and its price should reflect this.

Standard general conditions of contract may be used by the owner, but these have to be modified to eliminate reference to variations, latent conditions etc.

Incomplete *documentation* forces the contractor to estimate the missing parts and, unless misled or deceived by the owner, to allow accordingly.

The lack of a *latent conditions* clause forces the contractor to ascertain or estimate the actual conditions. Should actual conditions be different, the risk is taken by the contractor. The cost is with the contractor unless the owner misled or withheld appropriate information.

Associated *subcontractors* need to be aware also of the guaranteed maximum price and its implications.

Costs associated with *industrial actions* and indemnity from damages incurred by the owner could be expected to be transferred to the contractor.

Claims for *extensions of time* may be restricted to certain defined events, if allowed at all. The procedures to be followed by the contractor in making such claims could be specific and different to standard procedures.

Even if extensions of time may be permitted, *delay costs* would generally not be permitted.

Claims for *variations* may be restricted to certain defined items, if allowed at all. This forces the contractor to crystal ball gaze. The procedures to be followed by the contractor in making such claims could be specific and different to standard procedures.

Viewed from the owner's side, it would be unwise for an owner to accept an unrealistically low tender price. Rescission of the contract, litigation, arbitration or compromise may follow, as the contractor is unable to work with the low price. This will involve the owner in unwanted extra costs, and defeat the original purpose of having a guaranteed total cost.

Note that there can be a blurred distinction in many publications and discussions between GMP and the prime cost version – fee based on guaranteed maximum cost contracts.

3.4 PRIME COST CONTRACTS

Alternative names for prime cost contract (payment) types are *cost-plus, cost-reimbursement*, and *cost based*. Where incentives are involved, they might be referred to as *incentive contracts*. The term *do-and-charge* may be used in a prime cost sense, but also may overlap with schedule of rates thinking.

Under this arrangement, the owner reimburses the contractor for all work carried out in

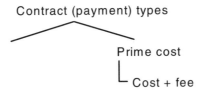

Figure 3.6. Categorisation of contract (payment) types.

connection with the contract including administration (project/field and business/general/ home-office/head-office overheads). In addition, the contractor is paid a *fee* for services, equivalent to a profit or a commission.

Sometimes the fee refers to both the contractor's administration and profit, sometimes the fee refers to only business overheads (not project overheads which are reimbursed) and profit, sometimes the fee refers to something else, and what actually constitutes the fee and what will be reimbursed needs to be clarified in any given situation.

Terminology

Prime cost
Prime cost is the net cost to the contractor of a particular item or piece of work.

Cost plus
Under such a contractual arrangement, the contractor is paid the prime cost plus something to cover overheads, profit, ...

Negotiated contracts
Some people refer to prime cost contracts as 'negotiated' contracts. However this terminology is not recommended because all contracts, no matter what their characteristics, can come about through either formal tendering or by negotiation.

General

Prime cost contracts are commonly used to cope with special problems, such as those encountered where:
• Owners desire to start construction/fabrication/erection without waiting for drawings and specifications to be developed to the point where competitive bids can be received.
• The project is of an unusual nature involving new techniques or experimental technology.
• The project is in a remote or relatively inaccessible geographic location.
• The project has other features making the risks involved difficult to appraise.
• The nature and extent of the work are difficult to determine before work commences; were the contractor to allow for everything, a lump sum price would be too high.
This type of contract can be used for any project work including design, documentation, procurement, installation, commissioning, construction and management. Minor work

(and daywork) can be arranged quickly with such contracts; actual hours and costs are tallied and reimbursed. Management service contracts are commonly cost-plus contracts, even though the head contract and subcontracts may be fixed price. Other examples include alterations and conversions, emergency work, fast-track project work, and highly complex or high risk work.

Daywork or *day labour*, whereby the contractor carries out work, generally not envisaged or priced at the time of tender but at the direction of the owner, is essentially done on a cost-plus basis, free of risk to the contractor. Variation work may be done on a daywork basis.

Where a contract is terminated part way through, completion of the work or to get the work restarted quickly, may only be feasible on a cost-plus basis. Quantum meruit ('as much as he deserved') payments to contractors may be cost plus in nature.

Although the fee will commonly be based on an estimate of the overall cost and duration of the work, the final cost to the owner is not usually known until the end. As such, prime cost contracts may not be favoured, for example, by public sector bodies that have budget constraints. Public sector bodies may also not favour prime cost contracts because of probity issues, including issues related to abuse of favouritism and justification of payments to contractors.

The estimated cost of the work is indicative but not binding.

Perceptions

In a perfect world, this type of contract would be a good way to tackle difficult projects. It puts the owner and contractor together so that they can act jointly (rather than in an adversarial manner) to produce the highest quality of workmanship possible, at the most economical cost. It also gives great freedom to adopt different methods of work, or to tackle unusual problems, or get out of unforeseen troubles. But individuals are not perfect, and before embarking on this sort of contract, the owner must assure itself that the contractor will undertake the work with an acceptable degree of efficiency, interest in the owner's goals and integrity.

Many owners and architect-engineers have the impression that the cost-plus type of construction contract is manna from heaven to the contractor. Actually, this may or may not be true. Many contractors prefer the lump-sum or unit-price form of contract where the profit is determined by the difference between the contract amount and the actual cost of carrying out the work. By dint of hard work and the exercise of his ingenuity, the contractor has the possibility of realizing higher percentage profits from [fixed price] contracts than he does from the usual cost-plus type. In contradistinction to this, however, the fee form of contract offers a certain profit. It represents a legitimate profit under circumstances where the contractor cannot lose. Industrial contractors, whose contracts are predominantly of this type, have achieved an enviable reputation for square dealing and integrity. The contractor must exercise particular care in such contracts to keep his dealings with the owner scrupulously honest and above board. Failure to do so will result in the loss of valued clientele.

(Clough, 1960)

Reimbursable costs

Common disputes in prime cost contracts are over what actually constitutes *reimbursable costs*. Accordingly, reimbursable costs should be defined explicitly in the contract along with the means of measurement of these costs. All other costs not so defined are the responsibility of the contractor (and would be expected to be covered by the contractor's fee). This also needs to be brought clearly to the attention of the contractor during tender negotiations, to avoid subsequent disputes.

Consideration needs to be given to such obvious things as:
- Labour (direct and indirect), subcontracts, consultants – on costs, accommodation, transport, expenses, insurance, fringe benefits.
- Materials – transport, insurance, loading charges, holding costs, duties and taxes.
- Equipment – transport, insurance, operation, maintenance, depreciation, duties and taxes, rentals, small tools (an agreed lump sum per worker).
- Services – water, power, fuel, postage, telephone, electricity; provision, maintenance.
- Health and safety related costs.
- Project facilities – warehouse, workshop, temporary buildings, temporary works, erection, transport, removal, office, furnishings, signage, cleaning, maintenance, insurance, rentals, access road.
- Inspection, testing, approvals.
- Insurance (those mentioned and not mentioned above – fire, vandalism, theft, public liability, property damage, ...); losses due to fire, extreme events, bad weather, and natural causes, but excluding those due to carelessness or neglect.
- Taxes (local, state and federal).
- Overheads (those mentioned and not mentioned above – project-based and others directly attributable to the project).

It is a difficult task to compile a completely satisfactory list of reimbursable overhead expenses. In recognition of this fact, one of the following two schemes is customarily used. If the job is of reasonably large extent, the contractor establishes a project field office, which performs all of the necessary office functions directly on the site. All expenses, including salaries, incurred by the project office are considered to be legitimate costs of operation. Where the project is smaller in size, it is more usual simply to increase the contractor's compensation by a percentage or fixed amount, so as to reimburse him a reasonable sum to cover his overhead and other indirect expenses.

Only costs directly and solely assignable to the project can be approved for payment. Under either of the two systems of overhead compensation discussed previously, no overhead costs from the contractor's central office are considered as reimbursable. These non-reimbursable costs include salaries of the [owner's] and the contractor's office staff, overhead or general expenses such as office rent or advertising, and interest on capital used. In general, all expenses incurred because of the fault or negligence of the contractor are his own responsibility.

(Clough, 1960)

It is the not-so-obvious items that escape initial consideration that cause the most disputes. The owner and contractor are not aware such items are present until they come across them during the project. From experience, there always seems to be some item that is overlooked at the time the owner and contractor enter into a contract; the question then arises during the project as to who is responsible for the cost of that item. There is often no easy way out of such a dilemma.

Prime cost contracts place a reasonably heavy administration burden (in terms of detail, time, personnel and cost of personnel) on the owner. The owner is required to check and audit all the contractor's claims for payment. Approvals may also be necessary before the contractor commits expenditure.

Other items for clarification

Before starting work, agreement would be necessary on items other than reimbursable costs, including:
• Trade and other discounts.
• The costs arising from some fault (mistakes, damage to the work) of the contractor; contractors could be expected to make good defects at their own expense – this work and cost needs to be separated from other work in progress and its cost; there may be a retention sum withheld equal to, say, 25% of the fee owing, to cater for defective work.
• What are the industry accepted practices applicable.
• Contractor's head office overheads versus project overheads.
• Interest on money.
• Industrial relations related issues.
• Inclement weather.
• Subcontracting – what work and proportion of work is allowed to be subcontracted; subcontract (payment) type, i.e. fixed price or prime cost.
• Method, valuation and frequency of payment to contractor; prescription as to how fee will change with changes in scope.
• The accounting methods to be followed; bookkeeping, purchasing and invoicing.
• Hourly costs of equipment, whether rented or owned by the contractor or purchased by the contractor on behalf of the owner (salvage value to the owner); possibly agreed ownership expenses (allowing for legitimate profit), with operating costs reimbursed; when rental costs become significant (say, 80% of the purchase price), a 'recapture clause' may allow the owner to purchase the equipment off the contractor, with the rental payments contributing to the purchase cost; treatment of consumables.
• Who undertakes the supervisory, management and inspection roles.
• Involvement of the owner in establishing the work method, including work approval and overtime issues.
• Lowest cost versus reasonable cost of materials, competitive quotations versus materials available when needed.
• Credit for surplus materials.
• Owner-supplied materials.
Commonly contentious or sensitive reimbursements are the contractor's administrative costs, costs relating to the award and control of subcontracts, and charges for equipment. Exactly what the owner is paying for in these areas needs to be carefully defined.

Scope of work – management contracts

For management contracts, some people suggest that the scope of work be only described very generally. From the owner's viewpoint, this avoids the possibility of the contractor claiming an increased fee. Such contracts may contain wording with the intent of the following:

- *The documents do not completely delineate all the services that the owner may request.*
- *The owner is not bound to give the contractor any particular work.*
- *The contractor is not entitled to any additional remuneration above that agreed, no matter what services are provided or what changes occur.*

(Tyrril, 1989)

This may seem a very general description but when an individual enters a contract of employment the description of the work is no more specific. The salary (the percentage in the management contract) must be fixed but the description of the work to be done can be in general terms. A basic problem with ... many agency arrangement management conditions of contracts is that they try to spell out everything which the [management] contractor must do. This course results in something being omitted or argument over the interpretation and leaves open the possibility of a claim by the [management] contractor for extra remuneration.

It also leaves open the possibility that if the quantity of work is less than anticipated, the [management] contractor could claim loss of profit which would have been earned had all the work been carried out.

(Uher & Davenport, 1998)

However, while this addresses the cap on the fee payable to the consultant, it does not address inadequate or irrelevant work or fruitless directions taken by the consultant, and for which the owner is liable to pay (unless such matters are somehow incorporated in the fee).

Fee calculation

There are various ways that the fee can be specified:
- As a percentage of the estimated or actual cost of the work.
- As a fixed fee.
- As a variable (sliding) percentage related to the actual cost of the work.
- As a fixed fee together with bonus and penalty provisions related to the actual cost and time of the work, compared with the estimated time and cost of the work.
- As a fee based on a guaranteed maximum cost of the work.

See Figure 3.7 for two of these fee types.

The fee may be specified by the owner, or it may be part of that tendered. Different fee types offer different incentives to the contractor.

Fixed percentage fee

Such contracts find a use in urgent, extra or minor work, in situations where, say, the de-

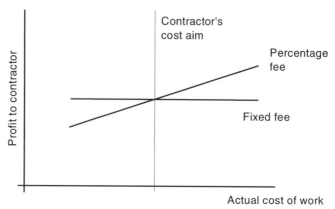

Figure 3.7. Prime cost contracts.

sign is proceeding side-by-side with the development, where the owner wants flexibility, or where the scope is indeterminate, and in emergency situations in order to get a contractor working as soon as possible.

Some example applications include war work, clean-up and repair after extreme natural weather events, remodelling work, and demolition work.

Contracts based on a fixed percentage offer no real incentive to the contractor to do the work at least cost or in least time. The more inefficient the process and the workers, the greater the contractor gets paid. Inefficiencies may also have a flow-on effect of low project (worker) morale. Conversely if the contractor puts in extra effort and reduces the project cost in the interests of the owner, this contractor works against itself by receiving a smaller fee.

The contractor may not like such contracts if it can do nothing without sanction, without detailing invoices, payment sheets, plant records and materials sheets etc., without auditing and without authorisation before receiving payment.

Cost plus percentage contracts may sow seeds of mistrust right from the beginning of the contract, while they saddle both parties with paperwork.

Some past abuses of cost plus percentage contracts have left the contract type with a bad public image. Some abuse has come from contractors not acting in the interest of the owner, and also not being content with a reasonable profit. They may have killed the goose that laid the golden egg.

The financial risk rests with the owner (see Fig. 3.7).

The acronym CPPF might be used to denote cost plus percentage fee.

Fixed fee

The fixed fee is usually based on a known broad scope of work, an initial broad estimate of the cost and an initial broad estimate of work duration. With a fixed fee there is no incentive to the contractor to delay the completion of the work, but against this, there is no incentive for the contractor to operate efficiently; in fact it may be in the contractor's interests (but not necessarily in the owner's interests) for the contractor to incur extra costs (e.g. expensive materials, expensive expediting) in return for earlier completion. The selection of the contractor is thus according to qualifications and not fee alone.

Although termed a 'fixed' fee, there may be provision for adjustment of the fee for changes in scope of work and changes in the prices of materials and labour.

The financial risk rests with the owner (see Fig. 3.7).

Variable (sliding) percentage fee

Contracts based on a variable percentage fee overcome some of the lack of incentives associated with the previous two prime cost contract types. Depending on the formula used to calculate the fee (but typically the fee decreases with increased project costs and times), there is an incentive for the contractor to finish on time and to a reasonable cost; there may be liquidated damages (damages applying for time overruns) as well as the percentage fee decreasing with the work going beyond a specified time. The work has to be sufficiently well defined at the start in order that realistic estimates for cost and time can be made, and certainly better defined than in a fixed fee arrangement There may be adjustments to the fee for changes in scope and prices.

Most of the financial risk rests with the owner.

An example (Miller, 1962; Antill, 1970) of the way the contractor's fee might be calculated is

$$\text{Fee} = R\,(2E - A)$$

where E = estimated (target) cost (excluding fee); A = actual cost (excluding fee); R = base percentage (rate) (either tendered by the contractor or nominated by the owner; the rate received by the contractor should the target cost and the actual cost turn out to be the same)

$$\text{Rate contractor receives} = 100 \times \text{Fee}/A$$

This is plotted in Figure 3.8 for a range of base rates.

Example		
Target project cost (E) of $100,000. Base rate ($R$) of 10%.		
Actual project cost (A), $	Fee $	Rate contractor receives, %
90,000	11,000	12.2
100,000	10,000	10
110,000	9,000	8.2

A contractor could be expected to know its approximate costs, and the amount it would expect from any project. Under such circumstances, the contractor works from its expected return to establish a bid rate, R. The above fee formula is used in such a calculation but with R on the left hand side.

Example
On a project where the owner has estimated a target project cost (E) of $100,000, a contractor expects the final cost (A) to be $115,000, and wants a return (Fee) of $12,500 to cover costs, profit etc. The contractor's bid rate becomes,
$R = \text{Fee} / (2E - A) = 12{,}500/(200{,}000 - 115{,}000) = 14.7\%$

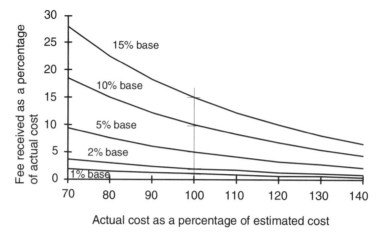

Figure 3.8. Curves for variable percentage fee.

Consultants are commonly paid on a sliding percentage scale. Professional bodies looking after the interests of their consultant members publish recommended scales of fees. For example, a design consultant may get paid a percentage of the project cost, with the percentage decreasing with the size of the project.

Fixed fee with bonus/penalty

Based on an estimate of the cost and time for the work, the contractor's fee is adjusted up or down depending on the actual cost and the actual time. For cost or time overruns, the fee is adjusted down (penalty); for actual cost or actual time less than estimated, the fee is adjusted up (bonus or reward). As with the variable percentage fee contract, there is a requirement for the scope of work to be adequately defined at the start, though changes to the fee with changes in scope and prices are usual.

The contract type provides an incentive for the contractor to work efficiently, while the owner is not penalised if the contractor cannot finish the work for the estimated cost.

Care has to be exercised in establishing the project cost estimate. Too conservative an estimate is too easy for the contractor to achieve; too low an estimate is too hard to achieve. Using a bill of quantities is one way of ensuring a reasonably reliable estimate.

Target estimate contracts, target cost contracts or *target price contracts* are such contracts.

Example

Examples (Antill, 1970) of the way the contractor's fee might be calculated are:

$$\text{Fee} = F\left(1 + \frac{T - A}{T}\right) \tag{1}$$

or

$$\text{Fee} = F + n\left(T - A\right) \tag{2}$$

Both give a fee reflecting any cost saving or over-run.

Here F = lump sum base fee (either tendered, or preset by the owner); T = target estimate of cost; A = actual cost (excluding fee); n = factor, e.g. 0.3, 0.6, distributing the cost savings or cost overrun between the owner and the contractor; where $n = 0.5$, the saving or overrun is split half-and-half (50:50) between the owner and contractor.

Tendering arrangements can have, in various combinations, the tenderer and/or the owner specifying the target cost and/or the sharing arrangement for cost savings and cost overruns, or the target cost and sharing arrangement can be negotiated between the contractor and the owner. Where tenderers are involved in bidding the target cost and/or sharing arrangement, competition is introduced.

The target cost may be subject to adjustment for variations.

There might also be an upper limit placed on the maximum fee payable, and a lower limit (e.g. zero) placed on the minimum fee payable.

Example

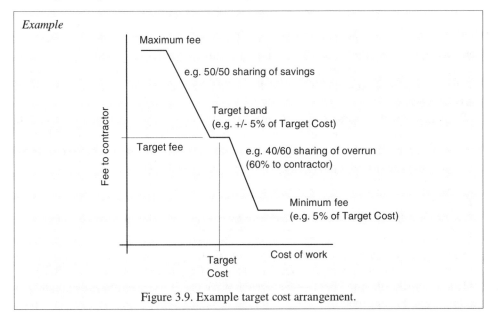

Figure 3.9. Example target cost arrangement.

For time savings, bonuses may be offered, for example, ranging from 50% to 100% of the associated penalty for time over-runs. Time bonuses and penalties may be applied independently of any cost bonuses and penalties.

It seems reasonable to give the contractor an incentive for being efficient, but in practice it may not always work. The main reason is as follows. If all the operations on a project can be specified, drawn, and quantified in advance, there may be no reason to have a cost-plus contract at all. A schedule of rates contract could be used instead. As well, it may be impossible to set a 'target' when all the operations cannot be foreseen, when the extent of the work required cannot be measured in advance, or when risks attached to the work are significant. The risks may or may not materialise.

A target contract can be a kind of contradiction. When a risk materialises which is other than a risk for which due allowance has been made in the target, or when an operation or an amount of work has to be done which was not included in the target, the target must be changed to be fair to the contractor. Contracts that involve extensive modifications to the target as the work proceeds, can be a source of friction and dispute.

A combined fixed fee (or percentage fee) and target cost contract can also be used. For example, the contractor's head office and profit are covered by a fixed or a percentage fee. The remainder of the contractor's fee is based on a target cost arrangement.

Fee based on guaranteed maximum cost

With the contractor paying for everything beyond the stated maximum ('upset') cost of the work, such risk is accordingly reflected in the size of the fee (a fixed amount). The more indefinite the scope of work is at the start, the higher will be the fee.

Alternatively, the contractor tenders both a guaranteed maximum cost and a fee. The total sum of guaranteed maximum cost plus fee could be expected to rise, the more indefinite the scope of work is. Hence, reasonably well-defined work is required at the tendering stage.

The guaranteed maximum cost figure may be based on a target estimate increased by a percentage, for example 5% or 10%, or it may be the target estimate itself.

The guaranteed maximum cost and the fee are varied with changes in design, scope, conditions and prices (escalation) and delays. This may negate the effect of a guaranteed maximum cost.

Savings below the guaranteed maximum cost may be split between the owner and the contractor (bonus), based on percentages specified by the owner, or as tendered by the contractor, or as agreed between the tenderer and the owner. However, for large under-runs, the owner may resent paying large amounts to the contractor.

Example
Cost saving splits of 25%-30% to the contractor have been used on high-rise office projects.

Such contracts represent a modification of the GMP (guaranteed maximum price) contracts discussed under fixed price contracts. The GMP contract could be argued to be possibly a better deal than the bonus-free version of fee based on guaranteed maximum cost contract type, for the contractor who is able to bring the contract in under the maximum price, although GMP contracts may not allow for changes as this prime cost version does. The prime cost version guarantees the contractor a return (provided the maximum cost is not exceeded). The 'guaranteed maximum cost' (in the prime cost version) is variable, whereas the 'guaranteed maximum price' (in the fixed price version) is not.

Note that there can be a blurred distinction in many publications and discussions between GMP and fee based on guaranteed maximum cost contract types.

The guaranteed maximum cost approach represents a blend of fixed price and cost reimbursement elements. Of the prime cost versions, the fee based on guaranteed maximum cost approach passes most risk to the contractor. The contractor may even be asked to bear risks that are not within its control.

Example – fee quanta
As mentioned, what the fee is intended to include, can be different from one project to the next,

and in any particular case needs to be clarified. Indirect project costs and head-office overheads can vary greatly from one project to the next.

Barrie & Paulson (1992) published some 1970s data on fees (Fig. 3.10), which the authors feel remained little changed through the 1980s. The fee included head-office overheads and profit; project/field costs were reimbursed. Approximately 50 projects were sampled, covering different delivery methods, different stakeholders, and both public-sector and private-sector work.

Barrie & Paulson (1992) also published some Department of Energy 1989 data (Fig. 3.11) on fees covering head-office overheads and profit for delivery methods like construction management.

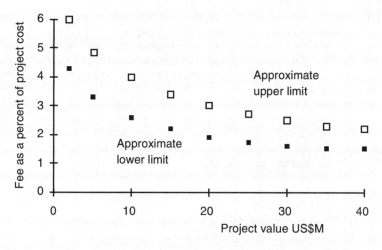

Figure 3.10. Approximate fee guidelines (Barrie & Paulson, 1992).

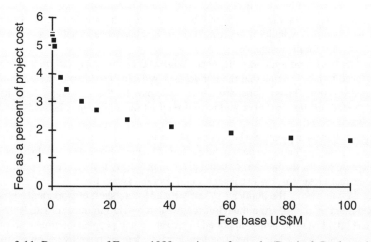

Figure 3.11. Department of Energy 1989 maximum fee scale (Barrie & Paulson, 1992).

Case example

A builder was engaged on a cost plus a percentage basis to do some extension work to an existing house. The agreement allowed for the reimbursement of costs to the builder for all labour, specialist equipment hired and materials. The builder's previous work was examined and previous owners interviewed as to the builder's workmanship. A number of issues, that had not been foreseen, arose during the execution of the contract:

- The builder made mistakes in setting out the work – in the location of the footings, in the pitch of the roof, in the window size openings allowed for, and numerous smaller matters. All of these work items had to be redone, which involved extra labour and materials, or accepted as they were. The issue that arose was – who pays for the additional labour and materials resulting from the mistakes?

- Material quantities were overestimated by the builder, leaving surplus materials at the end of the job. The issue that arose was – who pays for the surplus?

- The percentage fee was intended to cover, amongst other items, the builder's management of the work. However in the progress claims, management was listed as a labour item on top of which a percentage fee was added. Obviously at the time of signing the contract, the owner and builder had a different interpretation of the term 'labour'.

- The contract stated that the builder would provide usual builder's tools and equipment. The builder's tool kit turned out to be very lean and many pieces of standard builder's equipment had to be hired. Obviously at the time of signing the contract, the owner and builder had a different interpretation of the terms 'standard' and 'usual'. The issue that arose was – who pays for the hired 'standard' equipment?

- Days involving intermittent rain meant work started-stopped-started-stopped etc. The rain was not sufficient to send workers home. The issue that arose was – who pays for the unproductive time when it is raining?

- Planning by the builder was essentially a sequential mental process, namely 'we've finished this, now what do we do next?' Such nonexistent or poor planning by the builder was reflected in two outcomes. Firstly, time was wasted waiting for major materials to arrive. Secondly, continual trips and orders were made to the hardware suppliers for small items. The issue that arose was – who pays for the time wasted because of poor planning?

- At the time of signing the contract, it was assumed by the owner that the builder carried all the necessary public liability insurance, workers accident insurance etc and this would be taken care of in the contract fee. The builder carried no insurance, but the local authority insisted as a condition of building approval that relevant insurance be held. The issue that arose was – who pays for the insurance?

- The builder possessed a mobile telephone. Work was continuously interrupted in receiving calls related to other clients that the builder had, reducing productivity on some days by an estimated 10-20%. The issue that arose was – who pays for the reduced productivity? A telephone existed on site, and this could have been used for important calls, and a mobile telephone was not needed.

- The builder had an apprentice who was used to assist the builder, clean up etc. It was not uncommon for the apprentice to be standing around looking for something to do, fetching lunch and morning tea for the builder, or being given unnecessary tasks in order to fill in the day. The issue that arose was – who pays for the time of the apprentice?

- During the job, two of the builder's industry friends were short of work, and so the builder engaged them on the job even though it represented overkill in terms of site labour. At one stage one bricklayer was being supported by two and a half people labouring (carrying bricks and mixing mortar) for him, when one labourer was demonstrated by the owner to be more than adequate. The issue that arose was – who pays for this inefficient use of people?

Case example

A firm specialising in the organising and running of conferences was engaged on a contract which was on a cost plus a fixed fee basis. The fee was for the contracting firm's management of the whole conference. The cost reimbursable part included telephone calls, facsimiles, printing of stationery and brochures and similar matters.

The owner's and contractor's understanding of 'cost reimbursement' turned out to be different. For example, the owner assumed that the amount reimbursed to the contractor for a telephone call would be what the telephone company charged the contractor, and would be the amount shown on the telephone company's invoice to the contractor. On the other hand, the contractor assumed 'cost reimbursement' for a telephone call was the telephone company's invoiced cost multiplied by a factor of about three to cover the contractor's overheads. The owner assumed any overhead costs would be taken care of in the fixed fee.

For those items that were cost reimbursable, there also was a difference of opinion as to what were necessary activities versus what were frivolous activities. The owner's concept of value-for-money, for example in telemarketing, differed from that of the contractor's concept. The owner knew the industry, whereas the contractor only knew about running conferences. This situation was corrected, but only after unnecessary expense.

Summary – prime cost contracts

Owner's position – advantages

- Fast-tracking is possible; trade off possibly extra payment to contractor against time saving and associated cost savings.
- Useful in periods of high contractor demand; or situations where contractors won't bid fixed prices, e.g. large, complex and long-duration projects where material and people price fluctuations are difficult to forecast.
- Allows owner participation and input to the work, and approach of contractor; allows owner to manage and control rate of expenditure, and quality; allows contractor input to design.
- Does not require quantities or scope definition prior to commencement; can be let quickly, e.g. to exploit an emerging market opportunity, in emergencies, for a speedy project etc.
- Flexibility to make changes; flexibility to react to unforeseen conditions.
- Minimum adversary between owner and contractor (provided reimbursability is addressed precisely); no disputes over variations.
- Subcontracting possible; eliminates marginal subcontractors that may be present in lump sum contracts.

Owner's position – disadvantages

- May not be the most economical alternative, particularly in a competitive market.
- A necessary, increased owner involvement; arduous contract administration and monitoring role checking all dockets, wage sheets, delivery dockets, ...; owner to provide approvals, and inspection and audits of contractor's work; generate more paperwork than other contract (payment) types.

- Uncertainty over some items regarding reimbursability; possible disputes; careful definition of what is reimbursable/not reimbursable required.
- No guarantee that hours worked or materials used are genuine; disreputable contractors can abuse the arrangement.
- Contractor's mistakes and rework commonly at the owner's cost; contractors without skills and knowledge can spoil the arrangement.
- Changes may negate any bonus/penalty/guaranteed maximum cost.
- Price competition unsure.
- Of all the contract (payment) types, most risk is with the owner; risks carried by the contractor under other payment types transferred to owner (e.g. industrial problems, weather, suppliers' delays, ...).
- Final cost and time not known; cost and time blowouts possible.
- Incentive for contractor to complete and be efficient lacking in fixed and percentage fee arrangements; relies on the integrity of the contractor; no commercial pressure on contractor to be efficient.
- Usually no retention sum.

Contractor's position – advantages

- Low risk; eliminates risk inherent in fixed price contracts.
- Minimal supervisory staff.
- Payment and profit assured; no commercial pressure to be efficient.
- Minimal tendering cost; contractor paid for planning and estimating which would normally be part of the cost of tendering.
- Changes in work or work due to unforeseen difficulties reimbursed.
- Contractor's mistakes commonly at the owner's cost.
- Minimum adversary between owner and contractor (provided reimbursability is addressed precisely); no disputes over variations; potential for ongoing harmonious relationship and future business with owner.
- Head office overheads can be reduced by appropriately (over)staffing the project and being reimbursed for such staff time.
- Minimal equipment holdings can be supplemented by rental equipment whose cost is reimbursed.

Contractor's position – disadvantages

- Changes may cause difficulties in planning and control; longer period planning may not be possible; a reactionary style of management might develop.
- Large owner involvement; contractor may resent intrusion of owner on what might be regarded as conventional contractor territory.
- A more favourable risk/reward ratio may be available on familiar project work with a lump sum arrangement.
- Contractor's reputation may suffer because of delays or excessive costs which may not be due to the contractor's actions.

3.5 INCENTIVES/DISINCENTIVES

Incentive/disincentive schemes, whereby both owner and contractor benefit, are difficult to envisage for fixed price contracts, unless they are based on time or quality issues. In a schedule of rates contract, the extent of the work to be done is at the direction of the owner and not under the control of the contractor; to attach incentives/disincentives under such circumstances may not work.

A bonus and/or penalty can be included in prime cost contracts, provided suitable targets/performance indices (outputs, efficiencies, ...) are set. Interestingly, penalties seem to be used more commonly than bonuses in practice; people seem to prefer the 'stick' rather than the 'carrot'. People seem not to offer a bonus unless there are definite technical or financial reasons. Penalties are included where the owner would suffer a financial loss, or where the owner would not comply with obligations outside the contract, or in some cases in the belief that the contractor needs to be brought 'into line'.

Terminology

The terms 'incentive' and 'bonus' are used in an essentially interchangeable form.

The law may not allow contract wording involving penalties, and so to achieve the desired effect or result of a penalty, the contract wording may have to be carefully phrased such that a penalty becomes, perhaps, an unattained bonus, or similar, that is, a disincentive.

The disincentives/penalties referred to here reflect the result. Care of their wording in contracts may be necessary.

Extensions of time are a constant source of problems. ... a way around the problem may be to have two separate prices, one for the works if they reach practical completion by a certain date, another if they reach practical completion after that date. The law does not permit a contract to impose a penalty. Therefore it is not possible to validly provide that if the contractor fails to reach practical completion by a certain date then the contractor will forfeit $1 m. It is possible to validly provide that if the contractor does reach practical completion by that date, the [owner] will pay the contractor an extra $1 m. The final result is the same. The failure to achieve practical completion by the specified date will cost the contractor $1 m but the law is not concerned with the result ... only with the words used.

Liquidated damages of a certain amount per day can also be imposed. However, sometimes there is a particular date which is very important to achieve and a daily amount for liquidated damages does not serve the same purpose as a substantial bonus/penalty.

(Uher & Davenport, 1999)

Bonuses and penalties may be used differently between different industries. For example, performance bonuses are more commonly used in mechanical engineering rather than civil engineering.

Examples
A simple bonus scheme pays the contractor a stated sum of money for every day the project is

early, up to perhaps a maximum amount. Alternatively the bonus may be a percentage of any cost savings. Or a combination of time and cost bonuses may exist.

A bonus for early completion may be set equal to the value of liquidated damages (a genuine estimate of the costs to the owner for late completion).

In one managed contract, where the fee is a fixed fee, and the project cost reaches 125% of the budgeted cost, the fee will not apply to further services, but rather the owner will pay the manager a reasonable fee. This is to encourage the manager to continue in the interests of the project.

Fixed fee plus profit sharing – Additional to the fixed fee, the contractor gets 25% of the underrun of the target, but obviously gets nothing extra if the target is exceeded. 25% to 50% is a common range for the contractor to receive.

Where a bonus scheme exists in a contract, and the owner's actions prevent the contractor obtaining that bonus, their may be a claim for beach of contract, with the damages to the contractor being the lost bonus.

It follows that bonuses could only be expected to be used in contracts where the owner's interference would be small or non-existent.

For prime cost contracts, Figure 3.12 demonstrates one possible basis for an incentive/penalty formula. It requires identifying a range of costs between a minimum agreed cost or price and a maximum agreed cost. This range might be termed a Range of Incentive Effectiveness (RIE).

For actual costs less than the minimum agreed cost the fee is fixed. For actual costs above the maximum agreed cost the fee is 0.

Within the range of incentive effectiveness, the fee to the contractor is given by,

$$\text{Fee} = \text{Max fee} \qquad\qquad AC < MinAC$$

$$\text{Fee} = \frac{\text{Max fee}}{MaxAC - MinAC}(AC - MinAC) \quad MinAC < AC < MaxAC$$

$$\text{Fee} = 0 \qquad\qquad AC > MaxAC$$

Where Max fee = maximum fee contractor gets; AC = actual (final) cost of work; MinAC = minimum agreed cost; MaxAC = maximum agreed cost.

Some variants on this approach are:
- The formula need not be linear, but could be nonlinear or piecewise linear.
- The basis for the fee calculation for cost underruns can be different to that for cost overruns.
- Incentives/penalties can be wider than on cost alone; performance, time and quality can be included (Fig. 3.13).

This gives

$$\text{Fee} = \text{Fee (cost incentive/penalty)}$$
$$+ \text{Fee (performance incentive/penalty)}$$
$$+ \text{Fee (time incentive/penalty)}$$

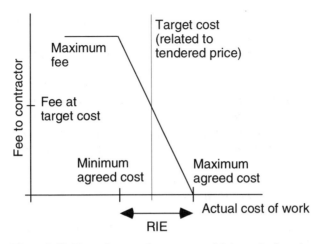

Figure 3.12. Example cost-plus contract with incentive/penalty.

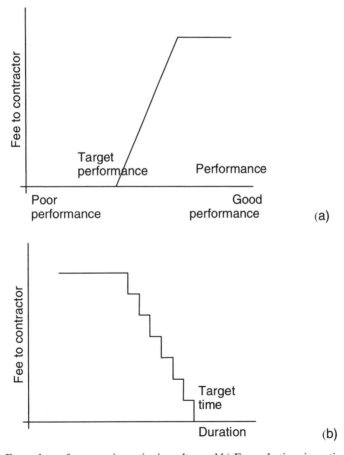

Figure 3.13. a) Example performance incentive/penalty, and b) Example time incentive/penalty.

... for high-technological projects with the manpower part in the order of 60% of the total cost, ... typical delivery incentives of the order of 1% per month of total contract costs. Different considerations were applied to penalties, leading to typical penalties (in the same environment) of 0.7% per month.

As each performance incentive, de facto, has to be tailored to the specific situation and characteristics, it is impossible to develop a general quantitative model.

As a general rule, it has been proven that the contractor will prefer pure performance incentives which are measurable before delivery (such as mass, power output, ...) to reliability incentives, which are only measurable after a defined period (such as ... maintainability, ...).

As for multiple incentives, only simple formulas have proven to be a workable tool and complex interrelated equations should be avoided.

An appropriate distribution of the incentive pool over the relevant elements should be carefully determined on a case-by-case basis. Any multiple incentive should be tailored to the specific [goals] of the project. A typical distribution of the incentive fee could be in the order of:
- *40% cost incentive*
- *30% performance incentive*
- *30% delivery incentive*

(Veld & Peeters, 1989)

Case example

A contractor on a large petrochemical plant in Africa was able to negotiate the following cost-plus incentive arrangement, because of a non-availability of suitable contractors to do the work. It was based on the estimated variability in completing projects of this type.

Total project cost allowing for risks: $1234 m
Total project cost not allowing for risks: $1123 m
Project cost difference: $ 111 m
Maximum fee to contractor (project cost difference): $111 m
Minimum fee to contractor (25% of project cost difference): $ 28 m
Sharing arrangement: 85/15 (see Fig. 3.14)

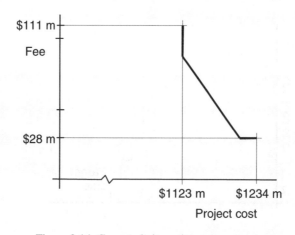

Figure 3.14. Case study incentive arrangement.

Case example

An incentive contract in the construction of a subway tunnel, 650 m long, with a finished internal diameter of 6 m. The soil condition was a combination of soft clay and fine sand.

Large liquidated damages applied for delays in completion.

The number of experienced workers available for shield tunnel construction was limited because the tunnelling technology was new to the country involved.

The tunnelling process repeated every one meter on a cycle – excavation, setting the reinforced concrete lining segments and grouting. The next cycle could not start until the previous cycle was completed.

Control of the work was essential – the settlement and rise of the ground had to be monitored, because this affected the excavation, speed and grouting volume. Less control could have resulted in collapse of the ground, concrete lining damage (cracking), and wrong alignment. Accurate and frequent surveys for alignment were required.

The cost of the machinery and equipment, including the tunnel boring machine (TBM), and the cost of materials, including the concrete lining segments, accounted for most of the project cost. The labour cost was not a significant part of the total cost.

In the first 150 m of tunnel, as a trial, the contractor was given a constant daily payment, irrespective of progress. At that point, the contractor and owner agreed that reasonable progress was 4 m per shift, and they agreed to a new incentive contract as below:

Progress (/shift)	Tunnelling labour ($)	Mechanical labour ($)
Up to 4m	A	1.2A
5 m/shift	1.3A	1.6A
6 m/shift	1.7A	2.05A
7 m/shift	2.2A	2.65A

$A/shift contained overheads, and was higher than the going wage. The owner had a say in manpower numbers, and in the direction and control of the workers. In a way, the arrangement between the owner and the workers was similar to a day labour system.

A fixed price contract was not considered appropriate because of the uncertainties. The completion date was also important. Effectively, the contract amounted to one of a fixed fee together with a bonus for earlier completion.

It would have been very difficult to get substitute tunnel workers should the contractor exit the project. As a result, the return to the contractor had to be guaranteed to prevent the contractor from leaving, should cost overruns occur. The provision of a penalty was not applicable for the same reason.

Performance and quality were totally controlled. Consequently, time was only the measure used to evaluate the bonus.

The contract offered a nonlinear bonus. The more the contractor performed, the more it got. It also avoided liquidated damages. There was little risk to the contractor.

To the owner, the impact of the extra cost caused by the bonus was not significant. It was offset in terms of reduced indirect costs.

Opportunities exist in most prime cost contracts for incentives. The incentives will take different forms, and will depend on factors such as existing practices, concern with meeting budget and schedule targets etc. Incentives are devised such that the contractor benefits from the owner achieving its particular goals. Contractors may be encouraged to make the project more profitable, work more efficiently or use better methods. For the contractor, an incentive is seen as a fair reward in return for giving the owner a desired outcome.

Incentives may be applied to all-sized projects, though they tend to be more common at the medium- to large-sized end of the spectrum.

Incentives may come about, for example, through the owner requiring an early start, or accelerated progress, or through conventional budget and schedule motives.

Incentives are devised to fit the specific project situation, and work through their being agreeable to both parties.

Gainshare/painshare

The terms gainshare/painshare might be used to describe a process of sharing of rewards and losses. Based on an agreed project target cost, there is a sharing of cost under-runs and cost over-runs according to an agreed formula.

An example formula is: the gainshare/painshare is allocated 50% to the owner and 50% to the other parties to the project, divided in proportion to each of the other parties' contribution. All win or lose together.

Figure 3.15 shows an example used on a resources project; the risk (financial exposure) carried by the participants had a cap at the level of their total gross margin.

The gainshare/painshare can be extended beyond cost considerations alone, to include reference to other project performance indicators.

The intent is to provide a strong motivational factor for all parties to work together, rather than in a confrontational or adversarial fashion, to produce a successful project. Performance and savings/profit are linked.

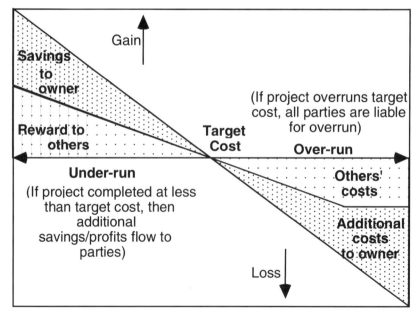

Figure 3.15. Gainsharing/painsharing diagram.

Case example
Smelter expansion – Performance incentive cost reimbursable contract

[The contractor] was selected to part design, manufacture, part supply, install, test and commission the alumina and bath handling system for [a] smelter expansion. The major elements ... were the construction of civil foundations, 14 large steel storage bins and associated steelworks, two air lift towers and associated galleries, installation of specialist equipment ... and associated electrical works.

[The contract payment type] was chosen for its suitability to this particular [situation]:
• Design not fully developed (fast-tracking required).
• Early delivery very important to the [owner].
• Some brownfield work.
• Flexibility of [owner] on price and delivery method.
• All stakeholders had a sophisticated view of risk management.

The major benefits ... were as follows:
• The project was completed in a 'world's best' construction time for this type of facility.
• Costs were well within budget.
• The contractor was encouraged to strive for continuous improvement.
• The contractor had less risk of losing money.
• The owner had the flexibility to change the scope of work.

To achieve optimum levels of project delivery success, the following performance incentives were set up:
• A 'fee modifier', based on monthly agreed scores on key performance indicators, was applied to the monthly claimed base fee, with a range of plus or minus 30%.
• The contractor shared 50% of the 'cost underrun' on the target estimate set up early in the project.
• A 'gainsharing' scheme was established to share the financial rewards for good performance. In [the contractor's] case it was distributed 50% to the workforce, 25% to staff and 25% to the company.

(based on ACA, 1999)

Maintenance contracts

Maintenance contracts might be based on lump sum, schedule of rates, cost-plus or *target cost with incentives*. For the last case, the target cost may be set equal to the previous year's recorded maintenance cost. Provided defined performance standards are met, the contractor is paid a fee which increases/decreases in line with this year's maintenance cost compared to last year's cost. For example, there may be a sharing of any saving over the previous year's cost. Here the target cost is used as a benchmark, rather than any guaranteed amount.

The target cost with incentives is intended to align the interests of the owner and the contractor, in that both gain by decreasing maintenance costs. Performance standards have to be enforced, else there could be a tendency to under-maintain.

3.6 CHOICE OF CONTRACT (PAYMENT) TYPE

The selection of the best contract (payment) type for any situation is difficult. Definitely the contract (payment) type should be chosen to match the particular situation, but general guidance rules are hard to find.

One suggested approach is to negotiate a compromise between the conflicting interests of owner and contractor, but this is not always possible.

A final decision on the choice of contract (payment) type could be expected to be influenced, among other things, by:

- Budget considerations.
- Cost uncertainty.
  – This can be influenced by guarantees, contractor reputation, payment currencies etc.
- Type of work/services, environment.
- Technical uncertainty.
  – This is more relevant for new technologies and contractors working in areas new to them.
- Availability of resources.
  – This refers to the resources of the contractor and the current workload of the contractor.
- Project duration, schedule importance.
  – Schedules that can realistically be achieved by a contractor will affect delivery incentives/disincentives.
- Performance importance.
  – As for schedule criticality.
- Contractor motives.
  – There are a number of reasons why contractors undertake work.
- Past performance and reputation of contractors.
- Delivery method.

However the choice of contract type may not be related to such criteria. Many owners request specific contract types (and delivery methods) often based on past successes (or failures) experienced by the owner or other organisation, rather than the above criteria. That is, the choice may be quite subjective in practice.

Example

The following example demonstrates Antill's preference for fixed price contracts (Antill, 1970).

The relative total cost to the owner is the most significant criterion in the selection of the type of contract to adopt. Consider a project for which the works cost, without mark-up, is estimated to be $4,500,000, and the cost of design (whether done before or during the contract) $180,000. The owner's site supervision costs for a Lump sum contract would total about $160,000, for a Schedule of rates contract about $180,000, and about $220,00 minimum for any type of Prime cost contract (except a Guaranteed maximum cost, where it might be $200,000).

Realistic bid prices for the various types of contract would be of the order shown at (A) in Table 3.1. A mark-up of 10% has been assumed for the Fixed price contracts, and conservative fees of the same order for all the Prime cost contracts, except the Guaranteed maximum cost where a fee of 20% was adopted; in addition, a contingency of $110,000 has been included in the Lump sum contract price as a provision against incorrect quantities etc. The estimated total cost to the owner under the various contracts is shown at (B) in Table 3.1.

Table 3.1. Comparison of owner's total costs under different types of contract. [Formulae (1) and (2) refer to the formulae given earlier under target estimate contracts (see Example: p. 48).

Type of contract	Lump sum	Shedule of rates	Cost plus fixed %
	($)	($)	($)
Estimated works cost	4,500,000	4,500,000	4,500,000
Mark-up or fee	560,000	450,000	450,000
A. Bid total	5,060,000	4,950,000	4,950,000
Owner's design and supervision	340,000	360,000	400,000
B. Estimated project cost	5,400,000	5,310,000	5,350,000
Actual works cost	(4,970,000)	(4,970,000)	4,970,000
Actual fee	–	–	497,000
C. Actual contract payments	5,330,000	5,220,000	5,467,000
Owner's design and supervision	340,000	360,000	400,000
D. Actual project cost	5,670,000	5,580,000	5,867,000
E. Actual extra cost of variations to owner	270,000	270,000	517,000

Type of contract	Cost plus fixed fee	Cost plus variable %	Target estimates Formula (1)
	($)	($)	($)
Estimated works cost	4,500,000	4,500,000	4,500,000
Mark-up or fee	450,000	450,000	450,000
A. Bid total	4,950,000	4,950,000	4,950,000
Owner's design and supervision	400,000	400,000	400,000
B. Estimated project cost	5,350,000	5,350,000	5,350,000
Actual works cost	4,970,000	4,970,000	4,970,000
Actual fee	477,000	457,000	431,100
C. Actual contract payments	5,447,000	5,427,000	5,401,100
Owner's design and supervision	400,000	400,000	400,000
D. Actual project cost	5,847,000	5,827,000	5,801,100
E. Actual extra cost of variations to owner	497,000	477,000	451,000

Type of contract	Target estimates Formula (2), $n = 0.5$	Guaranteed maximum cost	Management at 10% fee
	($)	($)	($)
Estimated works cost	4,500,000	4,500,000	5,130,000
Mark-up or fee	450,000	900,000	513,000
A. Bid total	4,950,000	5,400,000	5,643,000
Owner's design and supervision	400,000	380,000	150,000
B. Estimated project cost	5,350,000	5,780,000	5,793,000
Actual works cost	4,970,000	(4,970,000)	5,400,000
Actual fee	350,000	–	540,000
C. Actual contract payments	5,320,000	5,670,000	5,940,000
Owner's design and supervision	400,000	380,000	150,000
D. Actual project cost	5,720,000	6,050,000	6,090,000
E. Actual extra cost of variations to owner	370,000	270,000	297,000

Suppose now that in each case the construction contractor had underestimated the works cost by $200,000 (i.e. 4.5%), and that sundry variations during the work resulted in a net increase in value of $270,000 (6%), both suppositions being quite realistic in practice. After adjusting the original estimates on account of the variations, the total actual contract payments for the work in accordance with the different contract conditions would then be shown at C, and the total costs to the owner would be as at D in Table 3.1.

The extra cost actually payable by the owner, which should not equitably exceed the $270,000 valuation for the variations, will be as shown at E. The data at D and E of Table 3.1 provide an interesting and significant quantitative measure of the advisability of adoption of these various types of contract.

If the project had been designed and built under a Management contract on a basis of cost plus 10%, and if the actual construction work had been carried out under separate subcontracts to the Management contractor at the unit rates previously assumed for the Schedule of rates contract, the total cost to the owner would then have been estimated as follows:

		$
Design by management contractor		180,000
Estimated construction cost (unit price subcontracts)		4,950,000
		5,130,000
Fee for management contractor, 10%		513,000
Estimated total bid	(A)	5,643,000
Checking of design and site work by owner		150,000
Estimated project cost to owner	(B)	5,793,000

In the event, the actual cost to the owner would have been:

Design by management contractor		180,000
Actual construction cost (at schedule of rates)		5,220,000
		5,400,000
Fee to management contractor, 10%		540,000
Actual contract payments	(C)	5,940,000
Checking by owner		150,000
Actual project cost to owner	(D)	6,090,000

It should be observed that, had the construction work been carried out in any other way than by Unit price subcontracts, the actual cost would have been greater.

From Table 3.1 it can be seen that, if the owner had the time and facilities to carry out his own design (or had he engaged a consulting engineer) and then call a Fixed price contract, the actual total cost of the project would have been of the order of $5.6 to $5.7 million, and the additional cost for variations $270,000. On the other hand, had he resorted to Prime cost contracts, the cost would have ranged from $5.7 to $6.05 million, with extra costs for variations running from $270,000 to $517,000. With a Management contract, using Unit price subcontracts, the project cost would be almost $6.1 million.

The financial advantage of Fixed price contracts is obvious.

3.7 CASE STUDIES

3.7.1 CASE STUDY – SEWER RETICULATION PROJECT

The contractor entered into a schedule of rates contract with a public sector organisation to supply and install various sections of a regional sewer reticulation scheme. The contract sum was approximately $3 m.

Typically the nature of contracts for all the sewer reticulation projects being administered by that public sector organisation was of the schedule of rates type, with lump sum contracts being used more for the construction of the associated sewer pumping stations. The schedule of rates for the contract was divided into five main groups, these being:
1. Establishment/disestablishment.
2. Excavation.
3. Pipelaying.
4. Backfill.
5. Manholes.

Each of the pricing groups was further subdivided to form the individual unit price items based on pipe sizes, depth of excavation, type of backfill material etc. Site Establishment and disestablishment were incorporated as unit price items, each with a nominated quantity of one. These particular items had pricing restrictions applied to them, in that the contract would not allow a price for site establishment to be in excess of $15,000, and similarly the price for disestablishment was not to be below $10,000.

The contractor proceeded to let subcontracts for the excavation, laying, backfill and testing of the sewer lines, and supply agreements with suppliers for the supply of permanent materials.

Based on the quantities given in the schedule of rates, the tender price included a markup to cover the following costs, additional to the obtained subcontract rates and material supply rates.
1. Project supervision.
2. Site setup and running costs.
3. Insurances.
4. Dewatering.

The contractor administered the contract on-site with one project manager, who performed several functional positions including site quality officer, procurement officer, contracts officer and project supervisor.

The project position was reported each month by the project manager to the company management based upon the costs-to-date, outstanding expenditure and cost-to-complete for the quantities in the schedule of rates.

This method of reporting took place for the first 60% of the project and, accordingly, the tendered margin was being seen to be maintained. 60% of tendered quantities corresponded to six months of reporting.

It became apparent that whilst there was 40% of tender quantities yet to be completed, actual quantities indicated that the works were 85% complete. That is, the scope of work was less than that anticipated when tendering.

The net result of the difference between actual, and scheduled quantities, was a dramatic loss in forecast margin. There was a further reduction in the margin reported to-date due to there being no cost savings in the fixed operational costs as a result of the contract period not being fully utilised.

Due to the unforeseen reduction in quantities as a whole, what should have been a profitable project turned into a project which produced a small margin and highlighted problems in tendering and monitoring schedule of rates contracts.

EXERCISE

1. The experience with the contract demonstrated to the contractor that special care must be taken when tendering and administrating schedule of rates contracts.

   The tender documents included construction drawings detailing the longitudinal section of all the pipelines to be constructed. However these drawings were not consistent with the listed quantities in the tender documents.

   The longitudinal drawings provided all the necessary details required to perform a reconciliation with the scheduled quantities, except for unknown ground conditions, such as rock, for which there was a provisional quantity in the schedule as an 'extra and over' to the excavation.

   Should the estimator have made an attempt to correlate the scheduled quantities with other available information? If yes, why? If not, what other practices could the estimator have adopted?

2. Due to resource and time constraints, it is often assumed by estimators that the owner has, in the preparation of the tender documents, taken all care in determining the quantities, and as such the quantities should be reasonably accurate.

   In the case of this contract, there was no provision for amending rates; the owner had excluded the clause relating to 'limits of accuracy' for scheduled quantities. Accordingly, the reduced contract value resulting from a net decrease in quantities led to a loss of tendered margin without any grounds for claims off the owner. With only a single rate for each item being used, would a better tender have included rates variable with quantity?

3. Counter to this, the contractor can benefit when the owner has grossly understated the quantities and, provided the unit price has sufficient in-built margin, the tendered margins will increase. An obstacle to this form of 'windfall' is that there may be no provision to claim an extension of time against the resulting increase in the scope of work. Could a tender provide for a time extension, even though the resulting clause in the conditions of contract had been deleted?

4. A review of the project reveals that not only did the estimators not satisfactorily review the scheduled quantities, but they did not uniformly spread 'on-costs' and 'margin' across all the unit rates. Similarly the project controller adopted the same 'trust' in the scheduled quantities without fully reviewing the contract documents.

   Should work under a contract, which has excluded a 'limits of accuracy' provision, be estimated as if the nature of the contract was a lump sum type?

5. In this regard, to prevent the loss of the budgeted tender margin or to prevent a late completion date, there must be reasonable confidence as to the likely quantities.

   A contract which invokes small limits of accuracy, say 10%, can confidently be estimated without being totally confident in the scheduled quantities. However, should schedule items, which are heavily weighted with overhead and mark-ups to maximise cash flow, be revalued when outside the limits of accuracy?

6. On many large projects there is anecdotal evidence that some tenderers expend considerable resources during estimating to predict the likely outcomes of the project, with a view to increasing the contract value above the tender price. For example a contractor, while pricing a freeway construction, obtained an independent geological investigation and report of the proposed freeway route. The independent report indicated that the likely quantity of rock, which was to be paid as an 'extra and over', was far greater than the token quantity of rock mentioned in the tender

documents. The token quantity of rock was a provisional item which was allocated an arbitrary value for the purpose of obtaining tender prices. In its tender, the contractor submitted a large unit price for the removal of rock. The overall contract value was kept to the value it otherwise would have been, by sacrificing the value slightly of other unit rate items. The contractor won the contract and, in the absence of a 'limits of accuracy' clause proceeded to make considerable money from the excess rock. Express your view on the ethics of such an approach.

7. Whilst the sewer reticulation contract was small in comparison to the above freeway example, there is the underlying message that a detailed analysis of the tender documents, and any other readily available information, can hint to a clever manipulation of the schedule of rates to not only ensure that the tendered margins are maintained, but also that a respectable positive cash flow is achieved as early as possible in the project. Or, cannot all eventualities be catered for? That is, whatever unbalancing is carried out, can there always be a situation postulated where the contractor will lose money?

8. The sewer reticulation project was affected by poor foresight at tender time. However it was also affected by the owner withdrawing work, which at tender time all tenderers could have reasonably anticipated would be included. For example, there were prices included for turf restoration, including a heavy margin. The owner withdrew turf restoration from the contract. What lesson for the contractor can be learnt from this?

9. The project required both experience in estimating and foresight in reporting each month to minimise the drop in anticipated project profit. To report progressively a schedule of rates job, the management needs to be confident that the future work, which will generate reported completion profit, will actually exist. Should project managers be required to substantiate the 'quantities to complete' in the monthly report?

3.7.2 CASE STUDY – SUBSTATION DISMANTLING PROJECT

The contractor typically sought work from a newly formed 'state owned enterprise'; the types of contracts were usually lump sum with schedules covering various components of the work, and rates for variations as may occur.

Usually these types of contracts had clauses which tied payment to achieved milestones, and this placed constraints on the contractor's cash flow, due to an inability to substantially complete many sections of a project.

In the particular tender of this case study, the scope of works was broken into two clear components, with various pages of pricing schedules for each section of the work. The two sections were:

1. A 110 kV substation to be dismantled.
2. Dismantling and recovery of 105 km of dual 110 kV transmission line/circuit and all its hardware.

On this occasion the tendering officer had done an analysis of the requirements of the tender and had made a conscious decision to front-end load two components, in order to hopefully circumvent any negative cash flow effects that may be incurred. It was decided to cover both sections of the work to cover any eventuality; the tendering officer selected the dismantling of equipment on the substation in one section, and chose the dismantling of transmission towers in the second section of the schedule to load the price.

The tender was submitted. During the review period, the owner decided that there was an advantage if it took the work and split it into two separate contracts. Notices were sent to the appropriate companies tendering the work, to revise their respective prices and re-submit their prices as two separate tenders.

With some reservation the contractor's tendering officer did a quick split and resubmitted the separate bids.

Next, the contractor was requested to attend a tender clarification meeting on the line dismantling contract. The nominated project manager, with no knowledge of the scope of works, and the general manager drove the circuit, reviewed the methodology, had a discussion with one of the subcontractors and then attended the meeting.

There were the usual questions and answers on the program, methodology, location of bases, storage and handling and security, liabilities and subcontractor selection. The owner informed the contractor that the main subcontractor, on which the contractor had based the majority of the tender, was unknown to the owner, and so it would not accept the subcontracting company in that role, but would consider it in a lesser role.

The contractor left with a list of queries to respond to in writing, and to select another subcontractor. The contractor was given a day to consider and respond. At this stage the contractor did a quick analysis, obtained another subcontractor's rates for the work, and found with the increase in cost, it could still attain some margin, but the margin would be drastically reduced. The contractor agreed to undertake the work, feeling smug that it had been quick on its feet, and with some shrewd negotiation it could return some margin from the job.

A third of the way through the project, the owner dropped another 'bombshell'. The owner had, from day one, envisaged removing some of the transmission towers from the dismantling section of the contract, because it was negotiating to on-sell them as they stood to a local power authority.

A negative variation sheet was completed. As this part of the work was the front-loaded component of the contractor's tender, the contractor tried submitting a reduced value for that component of the works, mentioned parts of the works that it had already incurred costs, and used various other arguments. The contractor even tried the up-front approach, explaining that the tender had been front-end loaded because of the type of contract, and there was more in that section of the pricing schedule than the cost of the tower dismantling. This was all to no avail. Gone was any margin the contractor expected to make and the job was now starting to cost the contractor money.

EXERCISE
1. From the contractor's viewpoint, this project started to go wrong from the initial stages of tender and there were possibly two main reasons, plus a handful of lower level effects, that caused the inevitable loss.

   The two main perceived reasons were:
   (i) The tendering officer was possibly running short of time when doing the tender or didn't have a full grasp of the scope of work.

   The tendering officer had gone to the trouble of putting together an all-encompassing price, which was fine if the tender was let as one contract. Instead, the contract was broken down into two separate contracts; then again broken up after the contract was awarded.
   (ii) Because it was assumed by the tenderer that the risk of change within the scope of work was low, the pricing schedules were driven by trying to set up a mechanism to cover negative cashflow in the project.

   In what ways could you address these two areas of concern in future tenders?

2. The other flow-on changes that also had an effect were:
    (iii) For the subcontractor intended for a large component of recovery work, and who was subsequently rejected by the owner, more detail on the company's previous work history and suitability for that type of work should have been submitted in the tender and backed up with resumes of the relevant personnel that were going to be used specifically on that part of the work, or more work should have been undertaken to check their suitability prior to submission.
    In what other ways could this area of concern be addressed in future tenders?
3. (iv) The removal of various transmission towers. There was no hint during the pre-tender meeting of scope reduction. Usually if this was to happen, the owner would make a comment at the pre-tender meeting. Perhaps this was missed or the owner was unsure at this stage.
    In what way can such possible eventualities be catered for at the time of tendering?
4. In hindsight, this was a case of not enough time and possibly not enough understanding of the requirements of the contract. The other point is that usually time is the biggest enemy for tendering personnel.
    Are all the above tendering issues peculiar to the type of contract involved, or do they apply to other types of contracts as well?

3.7.3 CASE STUDY – BUILDING EXCAVATION CONTRACT

The project involved the construction of four buildings, one of which was the extension of an existing building. The total work was to be completed within a period of one year. A schedule of rates contract was used.

The existing building was constructed almost 15 years previously. The drawings showed that the footings of the proposed building were to be constructed adjacent to existing footings, with the space between being 2 m. When the excavation was carried out, it was found that the foundations of the existing building were almost 400 mm above the stated level and the level of the new footings. The alignment was also not exact. This resulted in the re-design of the footings and a new drawing was released after 16 days. The work in that portion of the building was delayed.

The new design was totally different from the original design; it required the excavation to be very close to the existing building, and it required heavy reinforcement and an increased concrete quantity. The area was very congested and hence an excavator could not be used. Excavation was carried out manually.

During excavation, it was also found that the existing backfilled material near the existing footings was Black Cotton Soil instead of Yellow Earth and was very loosely compacted. Hence, on many occasions the loose material just slid down though the barriers provided, creating many problems, and resulting in extra excavation. As per the contract, the Black Soil was to be removed from the site. Extra labour and trucks were employed; the work was further delayed, and the overall progress was greatly affected.

Because of the above events, the contractor demanded extra time and costs, arguing that the information provided by the owner was not sufficient and correct.

The owner replied that no extra money would be provided for the excavation beyond the given measurements and that the contractor was responsible for making the necessary arrangements for supporting the side material while excavating. And as far as the time was concerned, the owner agreed only to extend the time limit for the days the drawings were delayed because of the re-design.

This resulted in friction between the two parties, but because of the good relations between senior management of the owner and the contractor, it was agreed to settle the matter through a third party.

Third party decisions

The existing footings were not properly aligned and their level was not as shown in the drawings. It was the owner's responsibility to provide clear details of the existing building (constructed 15 years previously).

The backfilled material below the existing building was not the Yellow Earth as indicated in the contract documents. Its removal should be considered as an extra item because it was not included in the schedule of quantities. But, only the quantity that came within the given dimensions of the proposed footing should be measured, because it was the contractor's job to prevent the soil from collapsing while excavating.

The delay was due to the problems in the foundation of the existing building; it was not the fault of the present contractor and hence the contractor should be given a time extension.

EXERCISE
1. Many disputes arise out of unclear, imprecise, ambiguous or inconsistent documents, or parts of contracts that are not in writing. Does this really matter though if the contract is of the schedule of rates type?
2. In projects involving extensions of existing buildings, it is hard to be precise about what will be found. Is a schedule of rates contract the way to go for such work? What else does the owner need to incorporate into the contract?
3. A similar situation arises in cases where some materials are supplied by one subcontractor appointed by the owner, and the quality of the material is not of the required standard. Is a schedule of rates contract the way to go for such work? What else does the owner need to incorporate into the contract?

3.7.4 CASE STUDY – REDEVELOPMENT OF A LARGE CITY PARK

The contract types were selected to take account of the large cross section of works to be completed in a very short period of time.

Prior to deciding on the contract types, a work breakdown structure of the project was developed detailing the various work packages. From this, a schedule was developed. This process assisted the owner's project team transferring to paper the multitude of projects ahead and assisted in developing a solid understanding of the project requirements over the ensuing eighteen months. For the purposes of procurement the works were put in the following categories:
1. Civil work.
2. Building construction work.
3. Demolition.
4. Landscape and beautification work.

5. Services (power reticulation, sewerage, gas reticulation, communications and water reticulation).

Civil work; building construction work; demolition

Work over $1000

Expressions of interest were obtained from consultants and contractors. From these, shortlists were developed, and those shortlisted allowed to bid for various projects.

The contract type for work items 1, 2 and 3 above was in the form of *lump sum* contracts. The contract documentation included a schedule of rates which served both as a control mechanism on variations and a predetermined price for additional works if required.

Consultants included architects, engineers of various disciplines and project managers. All of the consultants were on a *fixed fee*, irrespective of extensions in time and project cost.

Both consultants and contractors fitted within the *traditional delivery method*, chosen largely due to government regulations, the project manager's previous experience, market forces and budgetary control.

Work under $1000

Contractors, which were specialists in their area to bid on a performance specification for the work, were selected. This work was in areas such as services provision. This was similar to a *lump sum, design-and-construct* contract arrangement. This approach removed a tier in the administration of the works and allowed the owner's project team to directly call for tenders, engage and manage contractors. A system was set up whereby the project team would oversee the development without being involved in the finer details. Specialist contract staff were added to meet needs as they arose. This method of staffing attracted people with specialist knowledge to fulfil short-term requirements.

Landscape and beautification work; services

Items 4 and 5 above were attended to through *piece rates*. The piece rate contractors required attributes such as attention to detail, the ability to take instructions and develop a solution, the ability to increase resources as required and a competitive *schedule of rates* to justify their selection. This was a difficult task, given the tendencies of piece rate contractors to stretch out the work requirements. This however was overcome by assigning the contractors many tasks to be completed within short timespans. If bulk work was available, they were more inclined to work efficiently rather than not.

EXERCISE

1. The major factors influencing the contract type and delivery method were: government regulations (owner requirements), time, cost, quality, project team resources, market forces and risk allocation. What other factors may have been involved?
2. The lump sum contract delivered several advantages to the project. It assisted the project team in defining the contents of the contract, inclusive of specification and drawings, and this helped in maintaining budgetary control. It assisted by allowing the project team to add contract special

conditions which assigned risk to the contractor. It also allowed the project team to take advantage of market forces and attain the best price for the defined work. A major advantage was that the owner lost the flexibility to make changes economically. This loss of flexibility was invaluable in achieving completion of the project in time. (In the past, senior management often acted on political pressures and tended to change the scope of projects. They did not understand the difficulties associated with change management.) The strategy did not allow for changes in scope without detrimental implications on the schedule. This is the sole reason why the larger projects were administered without too much difficulty.

Considering the bigger picture, is losing flexibility to change the scope of the project a disadvantage?

3. Administering contracts consumes resources within an organisation and can be viewed as a disadvantage. In this case, the owner appointed a project team to administer and direct the works. Experience was not available within the owner's organisation and the owner recognised the need for the team. The funds were available and the owner's operation was not disadvantaged by adopting a matrix type organisation. The owner chose to use a project specific unit which was disbanded at the end of the project. How else could the project have been administered?

4. The utilisation of piece rates was considered on its merits and adopted for the purpose of performing repetitive tasks in specialist areas. It afforded the advantages of direct control over the work, flexibility and the ability to react quickly. Management of the work consisted of assigning multiple tasks, with short deadlines, to the contractors, to avoid conflict and control costs. What disadvantages do you see in using piece rate contractors in such circumstances?

5. The disadvantages of not subjecting piece rate contractors to competition, and of having resource constraints, required careful management of tasks and scheduling. Other contractors were introduced into the arena frequently to perform defined tasks, and the efficiency and the capacity of each contractor was compared. This was done when the resources of the preferred contractor were exhausted. This system had the potential to work well by keeping the preferred contractor honest and allowing the project team to further test the market for other resources suitable to its needs. How else could competition be introduced?

6. The most significant benefit of the approach was that the team was capable of adopting other methods of delivery such as design-and-construct packages as the need arose. The team was in a position to fully understand the scope, and servicing of the budget, and act as a contract administrator.

A consistent disadvantage was overseeing consultants administer contracts on the team's behalf. For example, the building construction project managers made inexcusable errors in matters involving unions, and this impacted on the whole of the site. The civil design/management consultants seemed concerned with their own interests rather than servicing the owner's interests. Overseeing became very expensive and it was consuming too much of the team's time. The team decided to relieve the consultants of their duties and drew on the team's own abilities to manage projects directly, and design most of the remaining civil work component in house. This presented an added strain on the team, given its limited resources, and necessitated considerable overtime for prolonged periods. It did, however, keep with the team's vision of controlling the project totally. Is this approach recommended compared with, say, employing a contract design engineer to work as a member of the project team?

7. Piece rate work proved invaluable. Although difficult at the start, during the screening and trial process, the project ended up with some of the most competent contractors in the business. The major disadvantages experienced were that senior management tended to add to the team's work various urgent tasks, and the team often exceeded its resource capabilities. Other contractors were then introduced to perform certain tasks. This process kept the competition in the air and reduced complacency on those chosen for bulk works. A major disadvantage of this process was that it introduced rivalry amongst contractors. On the site there were up to sixty contractors

working for various groups and the project team constantly needed to manage this situation to prevent hostilities. This often necessitated rescheduling works to minimise interaction amongst contractors. What alternative approach might have been possible?

8. Contractor performance was reviewed throughout the project against performance indicators. This assisted the project manager in assessing the project position, and assessing whether the project team could cope. The procurement chosen was a determining factor in the proper functioning of the project. The procurement approach evolved to ensure that the project requirements were met. Do you agree, or is it not possible to change procurement approaches during a project?

3.7.5 CASE STUDY – CONSTRUCTION OF A SHOPPING CENTRE

The contractor was responsible for the design, project management and construction for the project. The design on the project was fast-tracked with the construction.

The contract was a prime cost contract with a guaranteed maximum cost and a share of savings agreement. The contract included a program. The preliminaries and supervision, design and project management, and off-site overheads were all lump sums with no share of savings. The subcontract costs (with rise-and-fall) were lump sums with a share of savings. The agreement was that the savings were shared equally, up to a maximum of 2% of the contract sum. Savings above 2% of the contract sum were returned to the owner.

Variations to the scope of work, requested by the owner or authorities, were valued by an external quantity surveyor engaged by the owner. Where the variation resulted in an increase in the contract sum, there was an allowance of 3% for overheads and 4% for profit.

Disputes were to be resolved by either expert determination or arbitration. Expert determination was to be used for disputes relating to date of practical completion, extensions of time, amount of any progress payment or valuation of variations. Independent expert(s), as required, were to be appointed by the president of the local arbitrators association. The decision of the expert(s) was to be final and binding. All other disputes were to be resolved by arbitration. Unless otherwise agreed by the parties, the arbitrator was again to be appointed by the president of the local arbitrators association. In this case, the parties could be represented by legal counsel.

The risks associated with extensions of time were the owners. The contract had an allowance of 15 days for delays beyond the control of the contractor. This included wet weather and national industrial action.

There was an acceleration clause which allowed the contractor to claim additional costs if the owner requested an earlier completion of the project. There was no bonus for early completion, and no allowance for damages due to late completion.

All progress payments were certified by an external quantity surveyor. Payments were to be processed by the owner within 20 days. The contractor was entitled to interest on monies not paid on time. Where a dispute arose, and the owner failed to pay amounts determined by the appointed expert, the contractor could suspend or terminate the contract, and was entitled to compensation for loss of profit and amounts payable to subcontractors.

EXERCISE

1. In this project, the development manager, project manager, design consultants and contractor were all part of the same company. The project was also not competitively tendered, but rather

negotiated, with the cost checked by an external quantity surveyor. What potential for conflicts of interest between contractual and company obligations is there in this situation?

2. The owner has a guaranteed maximum cost for the project with the potential for a share of savings. The contractor could, by design innovation, maximise profit, which was shared with the owner. Can you see any downsides to this arrangement? Can the analysis of the fairness of this contract for the owner and contractor be anything other than subjective?

3. Do you feel that disputes could arise over what are reimbursable costs in this case? Who takes responsibility for design, or construction mistakes?

4. The fee agreement for this contract was

$$Fee = F + n\,(T - A)$$

where

F = Lump sum base fee
T = Target estimate of cost
A = Actual cost (excluding fee)
n = 0.5 when $A < T$ and $n\,(T - A) < 0.02\,T$
n = 1.0 when $A > T$

The incentive for the contractor was to bring the contract in at least 4% under budget to receive the maximum fee. There was also a reason to overestimate the target sum by at least 4%. There was a large disincentive to the contractor if the target sum was exceeded. Do you believe that this formula gives adequate incentive and disincentive for the contractor?

5. The contractor took no risk associated with time penalties. However there were no time incentives either. The contract had an acceleration clause, but this was not used. The project finished ahead of time and with a share of savings in excess 4%. The contractor made the maximum fee and possibly made a profit on indirect costs, because the project took less time. Should agreement have been reached on the target estimates of cost and time between the owner and the contractor, prior to any mention of bonuses and penalties, on the basis that both the owner's and contractor's estimate of target cost and time would be influenced by optimising their personal returns?

6. In a prime cost contract there is the potential for the owner to pay for defective design or substandard work. There may be no protection in the contract for the owner against this practice. On this project there appeared a major defect in part of the roof, and this required the roof sheeting to be replaced. The contractor paid for the new roof because it was considered bad business to make the owner pay for the roof twice. Can prime cost contracts contain defective work clauses and defective design clauses like other contracts, and are they enforceable?

7. Are prime cost contracts with incentives appropriate for design-and-construct work?

3.7.6 CASE STUDY – A CONTRACTOR'S BUSINESS

The contractor, which is the centre of this case study, is involved in the management of design and construction. However, the projects are based on varying contract types and delivery methods. This case study highlights the contract types and delivery methods adopted in its business.

Contract types adopted by the contractor include guaranteed maximum cost, lump sum, and cost-plus fixed or percentage fee typically under design-and-construct, traditional, and construction management delivery methods, and are broken up as follows:

- Design-and-construct; traditional:
  - Guaranteed maximum cost (40% of work),
  - Lump sum (25% of work),
  - Cost-plus fixed or percentage fee (10% of work).
- Construction management:
  - Fee (25% of work).

Owners may or may not have some continuing involvement, and may or may not accept the risks associated with the projects, depending on the project situation.

The company favours the design-and-construct form.

Although the contractor is involved with competitive tendering, much work is won without tender, based on the success of previous projects and its marketing activities. The company promotes its strengths in managing projects to owners.

The contractor does not actively pursue competitive tendering, but will tender for some projects to keep/win back a relationship with an owner, for prestige projects and/or to remain competitive in the marketplace.

Some construction projects are fast-tracked. This works particularly well in design-and-construct projects, where the contractor has total control of all phases of the work.

Lump sum contracts often return the highest margins and provide some flexibility and opportunity. This is favoured where the scope of work has been well defined (particular where the contractor has been involved in the design work) and documentation is complete. The contractor is confident in its ability to perform to delivery schedules, with an acceptable level of quality and within budget. Although there is always a risk that profit may be negative, the contractor has the confidence to finish the project within project parameters at all times, thus maximising profits.

Extremely important to the contractor are the additional contract conditions covering dissemination of savings, either direct or capped under a guaranteed maximum cost contract. Some risks associated with additional costs resulting from incomplete documentation, nominated subcontractors, latent conditions, industrial relations, extensions of time, delay costs, variations etc are borne by the contractor. If the contractor's estimate is sound, the outcome results in a highly profitable project. Each guaranteed maximum cost contract stipulates shared savings which are either weighted in the owner's favour, or are weighted equally.

About one third of the guaranteed maximum cost contracts have capped savings, stipulating a ceiling for savings claimed. Anything additional to this goes to the owner. Alternatively, the owner first collects its portion of savings, and anything additional to that being distributed to the contractor. This removes any incentive for the contractor to negotiate with suppliers for better pricing on goods and services. Equally, any negotiations resulting in cost savings with suppliers erodes the profit on prime cost contracts (without bonuses). When negotiating procurement contracts with suppliers, prime cost contracts (without bonuses) and, to a lesser extent, capped savings on guaranteed maximum cost contracts, offer diminished opportunity to leverage purchasing.

EXERCISE
1. Some people suggest that owners are reluctant to commit to lump sum contracts because they lack trust and fear the contractor may gain an edge over them. They say it is more difficult to sell lump sum contracts where trust is an issue between the parties. However, the owner may engage

a quantity surveyor for an independent opinion to determine contract viability and a reasonable cost. What is your view on the relationship between trust and choice of contract type by the owner?

2. Sharing the savings can also have negative effects with owners. If the savings dividend is high, the owner's trust toward the contractor may diminish if it feels the contractor's tender estimate was inflated. Low savings, or only enough to satisfy the owner, can have negative effects on the contractor, where any incentive to buy better is removed. How do you balance the shared savings idea such that both parties are happy?

3.7.7 Case study – Local government – flooding and drainage

This case study describes the practices used within the flooding and drainage section of a rural council. The section consists of a section manager, a drainage engineer, four technical officers and supporting administrative staff. The section is supported by a design section and a construction section within the council. The main functions of the section are the investigation of flooding and drainage problems within the council area of over 1000 km^2, and preparation of flood and drainage management strategies to alleviate identified flood and drainage problems. Based on these strategies a plan of works is prepared, and this leads to drainage and flood mitigation projects.

The section's annual budget is about $5 million. This budget includes all the investigations and preparation of concept designs and detailed designs; and construction and implementation of the projects.

Due to a lack of resources within the flooding and drainage section, most of the major investigation and preparation of flood and drainage strategies, concept designs and detailed designs are prepared by external consultants. Some is done by internal staff. Construction work is carried out by contractors and by in-house resources. Work such as surveys and soil tests are done by contract.

Contracts for the preparation of concept designs are mostly fixed fee contracts, that is the consultant is paid on a time and expense basis to an approved upper limiting fee. Without valid reasons, this upper limiting fee would not be increased.

Contracts to prepare detailed designs are mostly fixed fee contracts.

Construction contracts depend on the type and size of the work. Commonly contracts are lump sum contracts and schedule of rates contracts.

Exercise

1. When awarding contracts, tenders/quotations from consultants are assessed based on upper limiting fees. Even though consultants submit their proposals on a time and expense basis, most of them claim their fees up to the upper limiting fee and are usually paid. There is no incentive for the consultant to complete the work early, because even if the work is completed early or late, the consultant would claim its fee up to the upper limiting fee. How does the council get around this?

2. If it is an investigation or a concept design, it is very difficult to write a technical brief covering all possible options and cases. During an investigation, some new ideas might arise, and if the consultant is asked to investigate the new ideas, the consultant would consider this as additional work and claim additional fees. When consultants submit their quotations/proposals, usual practice by the consultants is to allocate minimum time, but maximum rates of payment, for each of

their employees involved in the project. (The total tendered price = time × rate × number of employees.) Thus, when additional work is requested, they are able to claim at the maximum rate.

Example – For an investigation, proposals were called from five selected consultants. Council's estimated cost of the investigation from past experience was $60,000. Submitted quotations were: A = $31,000, B = $51,000, C = $56,000, D = $84,000 and E = $92,000. Council had a selection procedure based on multiple criteria, and not just price. When this was applied to the above proposals, B's proposal gave the highest value, with C = 2nd, D = 3rd, E = 4th, and A = 5th. However, when the allocated times for the project were considered, it was found that B had allocated almost half the time of C. That is, B's hourly rates were very high, and allocated time was low, to give a lower total fee to win the contract. From past experience, council anticipated some additional work during the investigation. Council considered that the time allocated by B was insufficient to perform the work, and B's proposal was rejected. Consultant C was commissioned for the project. Do you consider that this was a good decision aimed at minimising additional claims, where the proposed works were not well defined?

3. For drainage or flood investigation works, it is very difficult to estimate a target estimate of the cost of the work because, at the beginning of the investigation, there are unknown problems. It is therefore very difficult to include a cost incentive/penalty component within the fee structure. Even measuring a consultant's performance is very difficult. However, would it be possible to include a time incentive/penalty within the fee structure?

4. The council currently does not have any incentive/penalty component within the fee structure for consultants for flood and drainage investigations. However council has discussed these issues and would like to include some type of incentive/penalty. What would you recommend?

5. Most of the engineering consultants are very good on technical issues. However, when a political or social issue arises, they are very reluctant to advise council. To overcome this problem, committees and working groups are formed and these committees and working groups make decisions. Because these committees/working groups consist of politicians and members of the public, they always create additional work, especially in the case of developing a floodplain management plan. This work is normally not anticipated in an investigation, and therefore it becomes additional work. It is very difficult to avoid this type of additional work, or difficult to include in a technical brief to the consultant. How do you deal with this problem from a contractual viewpoint?

6. Progress reports, which are in detailed and summary forms, are necessary and these can be used to recreate the history of the job to resolve disputes. For investigation/concept design, consultants are required to submit a summary of progress to support their monthly claims. Because the nature of the work is such that you cannot see any physical progress, a summary of the progress is the only thing supporting a claim. These progress reports are very difficult to check. To make progress payments, the owner in most cases, has to believe the progress report. If the consultant reports that 30%, say, of the survey is completed, it is very difficult to check. However once a stage of work is completed, the owner can request the consultant to submit the details. How do you deal with this problem from a contractual viewpoint?

Some projects, in addition to having detailed and summary reporting procedures, may also carry out exception reporting, where unusual or significant events on site are reported, and some reasons as to why, for example, there was a deviation from the program.

3.8 EXERCISES

Exercise 1
What is your view on the ethics involved in unbalancing bids?

Exercise 2

How would you have resolved the issue in the case example involving the formwork contractor and the measurement basis for payment, assuming you also had not foreseen it at the time of entering into the contract?

Exercise 3

What value do you see in distinguishing between 'bill of quantities' contracts and 'schedule of rates' contracts. Answer this from the viewpoint of both the owner and the contractor.

Exercise 4

With cost-plus contracts (as compared with fixed price contracts), would the owner be more readily agreeable to accept the project as being complete before everything is done on the defects list? Which is the better option for the owner from a defects viewpoint?

Exercise 5

The situation in delineating a priori what is reimbursable/non-reimbursable in a cost plus contract is the same as that in risk management. One of the first steps in risk management is to carry out a risk event/source identification. Commonly this is done with checklists, brainstorming, questioning colleagues, using one's experience etc. But there is always the possibility that a risk event/source is overlooked, and hence no response plan is in place to deal with it.

What approach would you recommend, in order that the prime cost contract did not omit any potential costs in its description as to what was reimbursable and what was not reimbursable? (By doing this, you are in fact performing what people describe as risk source/event identification, but for a particular situation.)

Exercise 6

Consider the delivery of a material, organised by a management consultant. Two options are open:
- The owner could specify in detail that payment will only be made for the most efficient (cost-wise) delivery taking into account quantity discounts and minimum travel costs.
- The management contractor has no constraints on how it organises delivery. It is free to organise delivery in any quantity, at any time.

Which is the preferred option for the owner? Give reasons.

Exercise 7

A version of the bonus/penalty idea is to have a fee that alters in relation to a predetermined project cost in the following manner. For example, for a project carried out for less than the estimated cost, the contractor gets 20% of the cost savings; for a project carried out for more than the estimated cost, the contractor pays 10% of the cost overrun (Fig. E7).

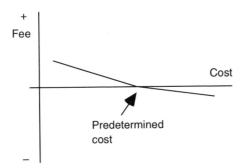

Figure E7.

What is your view on such an idea? How can time be included in the calculation?

Exercise 8
Some contracts for the sale of property give a commission to the sales agent based on a sliding scale percentage of the sale price; as the sale price increases, the commission percentage decreases (Fig. E8a). There is no real incentive for the agent to put in that little extra effort to sell the property, because the associated fee increase is minor, as shown in Figure E8a.

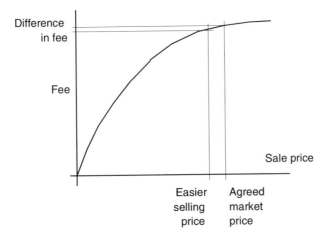

Figure E8a.

What would be a better way of calculating the sales commission in order that the sales agent gives maximum effort attempting to achieve the highest price for the seller? That is, how can the incentive for the agent to achieve a maximum sales price be increased and the disincentive to achieve any (lower) sales price also increased?

Would something like Figures E8b or E8c work?

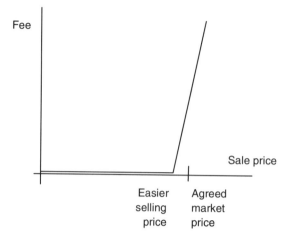

Figure E8b.

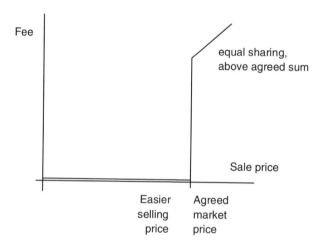

Figure E8c.

Exercise 9

One formula used for second-hand car salespeople is as follows. Each person receives a base wage of, say, $3000 per month, and the use of a car, whether cars are sold or not. For each car sold the salesperson receives a commission of $100 to $200 depending on the sale price (the higher the profit, the higher the commission), and for more than 20 cars sold per month receives a bonus of $2000 (Fig. E9a).

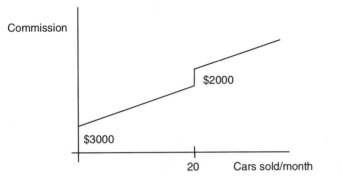

Figure E9a.

Presumably if no cars, or few cars, are sold the salesperson is out of work. Considering the relative values only of these money incentives, what is your view on this form of incentive and disincentive?

Would a commission of the form shown in Figure E9b be more in the interests of the salesperson's boss?

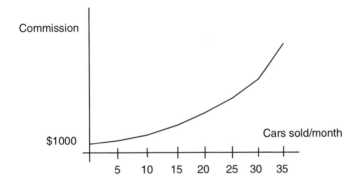

Figure E9b.

Exercise 10
How would you have resolved the issues in the house extension builder engaged on a cost plus a percentage basis case example, assuming you also had not foreseen them at the time of entering into the contract? In hindsight, these issues are easy to address, but how do we develop foresight?

Exercise 11
How would you have resolved the issues in the conference organiser case example, assuming you also had not foreseen them at the time of entering into the contract? In hindsight, these issues are easy to address, but how do we develop foresight?

Exercise 12
a) For the subway tunnel case example, the bonus was not distributed in its entirety to the workers. The company benefited from part of the bonus. How could the company justify this? What effect on worker morale might this have? What effect would TBM breakdown or maintenance have on progress and bonuses?

b) After the contractor become fully accustomed with the job, the owner' and the contractor agreed to the benchmark progress of 4 m/shift. In fact, this was very conservative, and the contractor achieved a considerable bonus over the total project. How might such a situation have been avoided by the owner, or is it of no concern how much bonus the contractor gets as long as the owner's goals are reached?

Exercise 13

With profit sharing below a target cost, would the owner be more readily agreeable to accept the project as being complete (at a cost below the target cost), or would the owner try to get everything done on the defects list up to and including the target cost? Which is the better option for the owner? Consider, for example, the prime cost case where the contractor gets 25% of the cost savings below the target cost.

Exercise 14

a) How might a project target cost be established? Presumably the project owner would negotiate for a lower target cost than the contractor.
b) Would it work if you made the project target cost part of what is tendered by contractors?

Exercise 15

List issues which you think would influence your choice in the selection of the contract (payment) type.

Exercise 16

Critically appraise the assumptions behind the example given by Antill on the choice of payment type. How realistic are the assumptions? How sensitive is the final result to changes in the assumptions made?

REFERENCES

Antill, J.M. 1970. *Civil Engineering Management*, Angus and Robertson.
Australian Constructors Association (ACA) 1999. *Relationship Contracting*, ACA, North Sydney.
Barrie, D.S. & Paulson, B.C. 1992. *Professional Construction Management*, McGraw-Hill.
Brooking on Building Contracts 1980. 2nd ed, by D.M. Bennett, Butterworths.
Clough, R.H. 1960. *Construction Contracting*, Wiley.
Miller, L.C. 1962. *Successful Management for Contractors*, McGraw-Hill.
NPWC/NBCC 1990. *No Dispute,* National Public Works Conference, National Building and Construction Council, Canberra.
The Aqua Group 1975. Which Builder?, Granada.
Tyrril, J. 1989. Project and Construction Management Agreement, *Australian Construction Law Newsletter*, Issue 2: 18-20.
Uher, T.E. & Davenport, P. 1998. *Fundamentals of Building Contract Management*, UNSW.
Veld. J. in 't & Peeters, W.A. 1989. Keeping Large Projects under Control: The Importance of Contract Type Selection. *Project Management*, 7(3): 155-162.

CHAPTER 4

Delivery methods

4.1 INTRODUCTION

Selection

There are numerous forms of delivery methods. There is no one best delivery method for all occasions. The more a delivery method is tailored for a particular project and players, the more chance it should have of being successful. Some common factors or constraints which influence the choice of delivery method include:

- Owner requirements; standardisation; political or organisational issues and policies; third party (e.g. financier) interests; corporate goals; project profile; probity requirements; accountability to others.
- Market conditions; economic climate.
- Time requirements; duration, deadlines, milestones; completion time and flexibility; timing and staging.
- Funding source and availability, cost requirements, budget and budget flexibility, cost ceilings; out-turn cost (final cost to owner); cashflow restrictions.
- Quality requirements; degree of quality control required by owner.
- Quantity/size of work; type/nature of the work; packaging of work; staging of work; technical complexity; degree of difficulty; greenfield or brownfield; new, refurbishment or maintenance; location; occupied or vacant.
- Degree of documentation; completeness and clarity of brief; design/ briefing time available; expected changes in requirements.
- Availability, abilities, experience, expertise and desire of outside and inside resources to do or administer the work, including the contracting and consulting industries; business acumen and financial substance of all parties; timing when outside consultants/ contractors are brought into the project.
- Availability of suitable conditions of contract (standard, modified or specially written) and contractual advice; experience of all parties with different delivery methods.
- Risks and allocation of risks; attitude to risk.
- Restraints – statutory, environmental, social, site, location.

Value management studies may help in selecting a preferred delivery method. There is no fail-safe selection process, because of the indefiniteness of circumstances surrounding projects. Each situation may require special tailoring.

There is no magic formula for selecting the best delivery method. The selection process is subjective. Some delivery methods rule themselves out because of their obvious unsuitability. However it may not be so obvious as to which is the best method.

Historical choice of delivery method

Over time, the popularity of different delivery methods has changed. The changes seem to be in response to changes in matters influencing project industries. These include changes over time in:
- Prosperity of the local, national and international economies (including inflation, interest rates, investment levels, ...).
- Local, national and international cycles in excessive and insufficient project work.
- Industrial peace/disputation; strength of unions.
- Addressing environmental (natural) concerns.
- An industry culture of large contracting/consulting organisations versus subcontracting/subconsulting.
- Asset/facility to be owned and operated, or used as an investment.
- Complexity and scale of projects.

For example, several decades ago, traditional, lump sum contracts were popular in the building industry. Typically the building was to be occupied by the owner, industrial harmony existed, subcontracting was not popular, and the economy was stable. Today's environment is different, and owners are searching for a preferred alternative delivery method, and this has given rise to a whole range of optional delivery methods. Today, perhaps the most pressing issue is the desire (for mainly economic reasons) to have the completed building as quickly as possible; this necessitates the commencement of construction before the design and documentation is complete.

Local, national and international cycles in excessive and insufficient project work have spawned periods of union strengthening/weakening. industrial disputation, favourable/onerous conditions of contract, and contractual claims and disputes.

Technology has improved and this enables faster-built and different facilities. Owner demands, such as increased services from the facility, have risen. But these don't appear to be the dominant driving forces behind the development of new or different delivery methods.

Return on investment stands out as a dominant driving force in determining new delivery methods. This includes recognition of the time value of money, disciplined management of all the project phases, minimal disruption to the project through disputes, and a finished product of appropriate quality.

The trend appears to be away from the so-called 'traditional' delivery method, though traditional contracting is still a major player.

Contractor organisations have steadily evolved from having departments of design, construction etc (functional groupings) and using a matrix arrangement to staff project teams, to almost exclusively being project-driven organisations with task force project teams that have a main allegiance to the projects rather than the individual, the organisation or the industry. By highlighting projects and not developing individuals, the organisation or the industry, this evolution may have longer-term detrimental consequences.

Choice of method

The selection of a preferred delivery method is influenced by the relative importance of many of the things mentioned above, as well as the financier's preference, or the owner's preference and past experience in the use of particular delivery methods. For example, the owner may have in place standard practices for a particular delivery method, and could be reluctant to try something different unless something goes wrong with the existing approach.

The delivery method, in part, establishes the business relationship between the contracting parties, and reflects (commercial) opportunities and risks to all parties. Each has differing characteristics relating to cost management, flexibility etc. Selection of delivery method can be seen as one aspect of risk management on projects.

Owners seem to prefer to have a single point of control in delivering projects. In highly complex projects, an independent third party manager may be employed to manage the project on behalf of the owner.

The right choice of delivery method is not the only issue that affects the success and performance of a project, but it is an important issue. Major projects always involve uncertainties and require continuous monitoring of every aspect of performance and prompt responses to changing conditions, irrespective of procurement policy.

Project development

Commonly, projects are planned at a 'macro' level first, well in advance of any consideration of a delivery method, and conditions of contract including contract payment type. The planning at this macro level sets the constraints and criteria for the later choice of any delivery method, and conditions of contract including contract payment type. Such an approach may be suboptimum, but does permit an ordered approach capable of being followed by project personnel.

Another suboptimum approach is one that chooses the delivery method, and conditions of contract including contract payment type, as separate entities, rather than as a whole package.

Generally, it is felt that the delivery method should be chosen early in a project's life. As the project progresses, the pool of candidate delivery methods could be expected to reduce and the optimum delivery method may be unfortunately constrained from being further considered.

The following discussion only applies to delivery methods for outside sources to do work, provide services or design, manufacture and/or supply equipment.

Acknowledgement

The No Disputes document published by NPWC/NBCC (1990) has been used extensively in the following. A number of passages are taken verbatim from this reference, while other passages have been modified or added to. The reliance of these notes on No Disputes reflects the opinion of this document held by the writer.

Table 4.1. Popular names for some delivery methods.

Item outsourced	Popular name
Construction ...	Traditional
	Construct only
Design (part), construction ...	Design novation
	(Novated design-and-construct; Design, novate and construct – DN&C)
	Detail(ed) design-and-construct(ion)
Design (part to full), construction ...	Design-and-construct(ion) (D&C)
	(also termed design-construct, design-and-build, design-build)
	(Performance contract; Package deal)
	Design-develop(ment)-construct(ion) (DD&C)
	Design-document-construct(ion)
	(Document-and-construct(ion))
	Turn-key
	Managing contractor
	Design-manage
Design and management	EPCM (engineering, procurement, construction and management)
	(Engineer procure construct)
Design, construction, ..., maintenance	Design-construct-maintain (DCM)
	Design-construct-and-maintain
	Design-construct-operate-maintain
Management (project)	Project management
	(Management contracting)
	Program management (multiple projects)
	Integrated contract (project management and construction management)
Management (construction, ...)	Construction management
	Professional construction management
	Construction project management
	Owner-builder
	(Management contracting)
Services (other)	... service
Supply	Supply
Design and supply	Design and supply
Many items commonly including finance	Commercial development
	Concessional methods
	BOOT, BO, BOT, BOO
	(build, own, operate, transfer)
	BOLT, BLO (build, operate, lease, transfer)

4.2 OUTSOURCING OPTIONS

Outsourcing of parts of a project and the completed facility may be with respect to:
- Finance,
- Design,
- Supply,
- Management,
- Construction (building)/fabrication/manufacturing,
- Operation,
- Maintenance,
- 'Owning' (over different franchise periods or indefinitely); leasing, and
- Combinations of these.

The way these are outsourced (delivery methods) might be called different names. Table 4.1 shows some conventionally accepted names for some of these delivery methods. Many delivery methods do not have popularly used names. Each may be practised with some flexibility, such that there is overlap between them, and *hybrids*, and the implemented forms may not fit one particular categorisation exactly.

Figure 4.1 might be useful for broadly distinguishing between some of the possible delivery methods.

More complete information is given in Table 4.2.

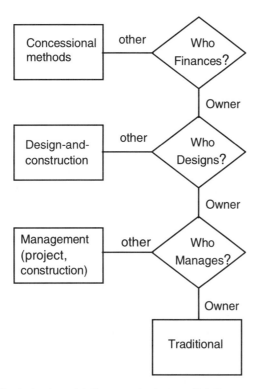

Figure 4.1. Broad discrimination of delivery methods (not all delivery methods are shown).

Table 4.2. Comparison of different delivery methods.

	Finance	Design	Manage	Construct	Operate	Maintain
Traditional				*		
Detail D&C Design novation		Part*	*	*		
D&C Turn-key EPCM Managing contractor		*	*	*		
DCM		*	*	*	*	
Project management Construction manage- ment			*			
Concessional methods	*	*	*	*	*	*

*Items outsourced.

Usage

Figure 4.2 is based on ACA (1999).

Early involvement

It is commonly argued that involving the key parties early in a project's life maximises the potential for effecting optimum project outcomes. Figures such as Figure 4.3 are often quoted illustrating the benefit of involving key stakeholders early in a project, where the ability to influence the final cost of the project or add value is greatest. (How such diagrams are obtained, however, is never explained.) Planning and design work carried out in the early stages has the potential to influence cost more than work, such as construction, in the implementation phase. Figure 4.3 also shows a 'cost of change' curve indicating that, if any changes are to be made, they should ideally be made early in a project. Alternatives should be examined early on.

ACA argues that the 'cost to change' curve is flatter where the contracting parties are working together, as opposed to an adversarial relationship.

Studies have demonstrated that 70% of the total life cycle cost (capital and operation costs for whole of life) are locked in by the time the project concept is resolved.
(New South Wales Public Works, 1993) (Fig. 4.4)

Collectively, the diagrams point to deciding on the delivery method as early as possible in a project.

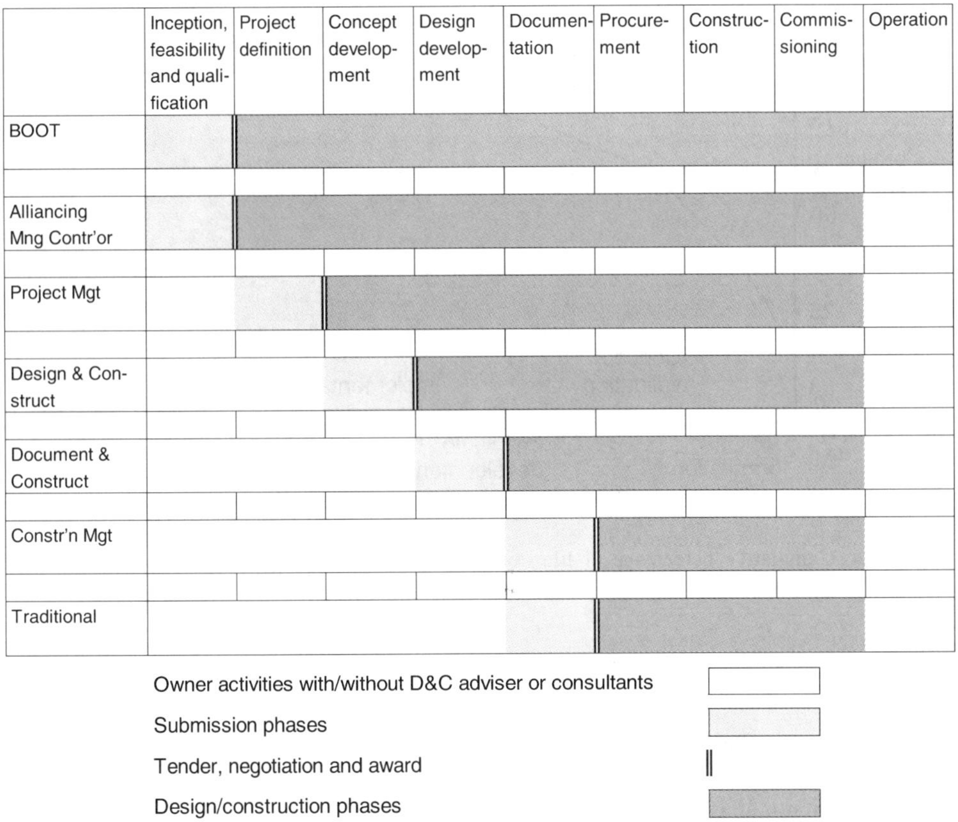

Inception, feasibility and qualification
Project definition
Concept development
Design development
Documentation
Procurement
Construction
Commissioning
Operation
BOOT
Alliancing Mng Contr'or
Project Mgt
Design & Construct
Document & Construct
Constr'n Mgt
Traditional
Owner activities with/without D&C adviser or consultants
Submission phases
Tender, negotiation and award
Design/construction phases

Figure 4.2. Alternative delivery methods (after ACA, 1999).

Concept
Development
Implementation
Termination
High
Ability to influence cost
Low
Ability to influence cost or Savings potential
Cost expenditure
Cost of change
High
Cost of change
Low
Time
Project start
Project end

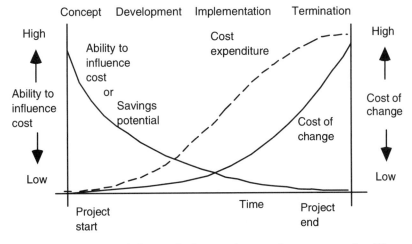

Figure 4.3. Ability to influence final cost and cost to change, over project life.

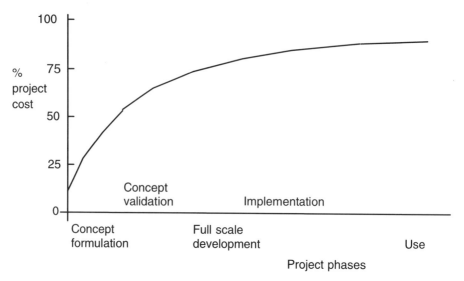

Figure 4.4. Cumulative percentage of life cycle cost affected (New South Wales Public Works, 1993).

Management contracts

The main forms of delivery using management contracts are:
• Project management.
• Construction management.
• Managing contractor.

Some people may make no distinction between the delivery methods. They may be all classed as *management contracting*, construction management or project management. (As well, the terms 'project management' and 'construction management' may be reserved by some people to solely refer to processes of management, rather than specific delivery methods.)

Also, other delivery methods may be called management contracts by some people, but there, management is not the basic obligation. Hence this is not a recommended use of the terminology.

With project management and construction management, the owner does not need to maintain full-time project management and construction management staff, but rather engages a manager on a needs basis.

Project management and construction management may be referred to as *agency arrangements*, while managing contractor as a *non-agency arrangement*. Agency arrangements could be described as pure management contracts. All the consultant's/contractor's actions are done in the name of the owner. The contractor assumes no financial liability. Non-agency arrangements keep the contractor status, but have the contractor perform only a managerial role.

All arrangements let the owner take most of the risk and result in a non-adversarial relationship between the manager and the owner. Payment of the manager on a cost-plus basis defuses most of the remaining conflict between manager and owner. Liquidated damages and completion dates, if present, tend to be hard to enforce, because of the nature of management contracts.

The problem with enforcing the time covenant in a management contract is that at the time the contract was let, the scope of the work has not been fully defined. A covenant to do an unlimited quantity of work within a fixed time is not legally binding. In essence, it is the same principle of law that prevents there being a lump sum price for management contracts. A contractor cannot be held to a lump sum where the work to be done for the lump sum is unlimited.

(Uher & Davenport, 1998)

Management contracts reflect the owner's desire to have some involvement in the project delivery. They represent a 'team approach' between the owner and the manager, non-adversarial in the interests of the owner's goals.

Management contracts represent the largest risk to the owner, and the least risk to the construction/project manager (assuming no guaranteed maximum price). The owner hopes, through effective and efficient management practices, that this risk to the owner is minimised.

The main argument advanced by the non-users of management contracts appears to be that there is a reluctance to use outside consultants as managers because control (particularly, financial control) wants to be retained in-house. There is a concern, for example, that the manager will commit the owner to additional expenditure, vary a contract, waive a right of the owner, accept defective work, permit omissions, or generally exceed its *limits of authority*. There can also be a concern that the manager will not maintain *confidentiality*, for example project information and records, such as accounting records, being disclosed to the media or others.

Generally, there is no compulsion for the construction manager or project manager to complete the project within budget or a scheduled timeframe, other than loss of reputation should it not do so. The contract with the construction manager or project manager is based on trust that the manager will act as a professional. The owner may also terminate the manager's services part way through a project should the project not be on track; the owner could also sue the manager for something like misconduct. Both of these options appear to be rarely carried out in practice. Individual contracts that the owner has with other contractors may include liquidated damages for late completion. Time extension clauses are only relevant in these contracts, and not in the contracts with the construction manager or project manager.

In construction management and project management contracts it is common for the [owner] to take out insurance for the works and public liability insurance in the names of the owner, the managing contractor and all consultants, subcontractors and employees. The insurance is then said to be '[Owner] Controlled Insurance'. If the construction manager is providing design services, the construction manager may be required to effect Professional Indemnity insurance.

(Uher & Davenport, 1998)

Other terms

Other currently popular procurement terms include:
* Partnering, alliances, relationship contracting.
* Fast-track.
* Period contract.

Partnering establishes an overlaying charter on any contract, whereby the parties to the contract agree on cooperation, trust, goodwill, the early resolution of disputes and a number of other non-adversarial measures.

An *alliance* or *relationship contracting* establishes cooperative behaviour between the parties to a project, a joint overseeing body, sharing of any gain or loss, and other measures attempting to have all parties working towards the same project goals.

Fast-track refers to the overlapping of a project's phases.

A *period contract* or *term contract* refers to a contract for continuing work or supply over a defined time period, rather than to projects. Alteration and repair work, for example, may be done on such a basis.

Which method?

Different projects with different owners and circumstances generally mean that there is no one best delivery method for all occasions.

Opinion is that a particular method should be tailored to the owner's and project's unique needs. Standard delivery methods may need to be modified.

The traditional method finds application where the owner is prepared to manage the interface between design/documentation and construction, and is prepared to exercise project control with respect to cost, consultants, construction and quality. Historically, the traditional method has been favoured in the majority of work.

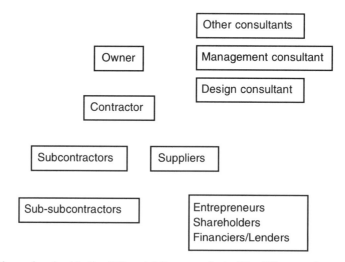

Figure 4.5. Players involved in the different delivery methods. The differences between the various methods is one of contractual and administrative links (lines) between the various boxes.

Lesser owner involvement is obtained in the detail design-and-construction method. The owner develops the concept design but has others take the project further.

Where the owner can prepare a brief detailing its goals, then a contractor can be employed to develop a suitable concept design, detailed design and carry out the construction as in the design-and-construction method.

Construction management may be applicable where the owner is prepared to manage the interface between the consultants and the constructors as well as the cash flow.

Some guidelines as to when each method might be used are given below.

Irrespective of which delivery method is employed, the same people/organisations and skills are employed (Fig. 4.5). The differences between the various methods is one of contractual and administrative links, the way different participants are coordinated, and the risk sharing between the participants. In simple terms, drawing the methods in terms of block diagrams with connections between boxes being the contractual and administrative links, then each method is a different shuffling of boxes and links between boxes, with the people/ organisations in the boxes 'wearing different hats' (different roles and responsibilities) for each shuffle (Fig. 4.5).

Suitability of different delivery methods

ACA (1999) offers the following approach to selecting the most appropriate delivery method. Different owners and contractors could be expected, while following a similar approach, to emphasise different criteria, and perhaps other criteria. ACA argues that both owner and contractor are best served if the delivery method, that best suits the project requirements, is chosen.

Figure 4.6 shows a suitability matrix, completed for an example project. Weights for each of the considered criteria are subjectively chosen. The project is then given a subjec-

Criterion	Weight [2]	Rating (1-10) [3]	Score = [2] × [3]
1. Is early delivery of end-product of value to owner?	20%	9	1.80
2. Nature of work – greenfield versus brownfield?	15%	8	1.20
3. Technology – proven or radical?	10%	7	0.70
4. Risk culture of owner?	10%	8	0.80
5. Tight guaranteed maximum price (GMP) essential for project sanction?	10%	7	0.70
6. Industrial relations environment?	10%	9	0.90
7. Proven relationship contracting record with potential engineering contractors?	8%	8	0.64
8. Sensitivity to disruption from aboriginal/heritage/environmental issues?	7%	7	0.49
9. Owner's understanding/experience of project delivery process	5%	5	0.25
10. Will construction require single (multi-discipline) or many contractors?	5%	4	0.20
Total	100%		7.68

Figure 4.6. Suitability matrix, example project (after ACA, 1999).

tive rating against each of these criteria. The product of columns 2 and 3 gives the score in column 4.

The rating scales for each of the criteria follow Figure 4.7.

Having evaluated a total score in Figure 4.6, ACA uses Figure 4.8 to select the most appropriate delivery method.

Risk allocation

In deciding on the risk allocation between contractor and owner, consideration needs to be given to not only the delivery method, but also the conditions of contract including the payment type. That is, risk can be considered at several levels. Something along the lines of Figures 4.9a, 4.9b and 4.9c is useful as a macro view in terms of control to the owner versus risk to the contractor.

Issue matrix

The New South Wales Public Works Department (1993) gives the matrix of Figure 4.10 to assist in the assessment of risks for several delivery methods. The figure lists the major issues that might affect an owner, and assists in the selection of the most appropriate delivery method.

Criterion	Rating 1 (Low rating)	Rating 10 (High rating)
1. Is early delivery of end-product of value to owner?	No value at all	Of great value
2. Nature of work -greenfield versus brownfield?	Total greenfield site	Many critical interfaces with existing operating facilities
3. Technology – proven or radical?	Well proven stable technology (will not evolve during project)	New and/or evolving technology
4. Risk culture of owner?	Totally risk averse – risk transfer culture	Strategic management of risk – sophisticated view of risk
5. Tight guaranteed maximum price (GMP) essential for project sanction?	Tight GMP essential	Owner flexible within range
6. Industrial relations environment?	Very low risk	Very high risk
7. Proven relationship contracting record with potential engineering contractors?	No track record or bad track record	Good track record
8. Sensitivity to disruption from aboriginal/heritage/ environmental issues?	Very low risk	Very high risk
9. Owner's understanding/experience of project delivery process	Little experience	Very experienced
10. Will construction require single (multi-discipline) or many contractors?	Will require many different contractors	Could be constructed by one contractor

Figure 4.7. Rating scales (after ACA, 1999).

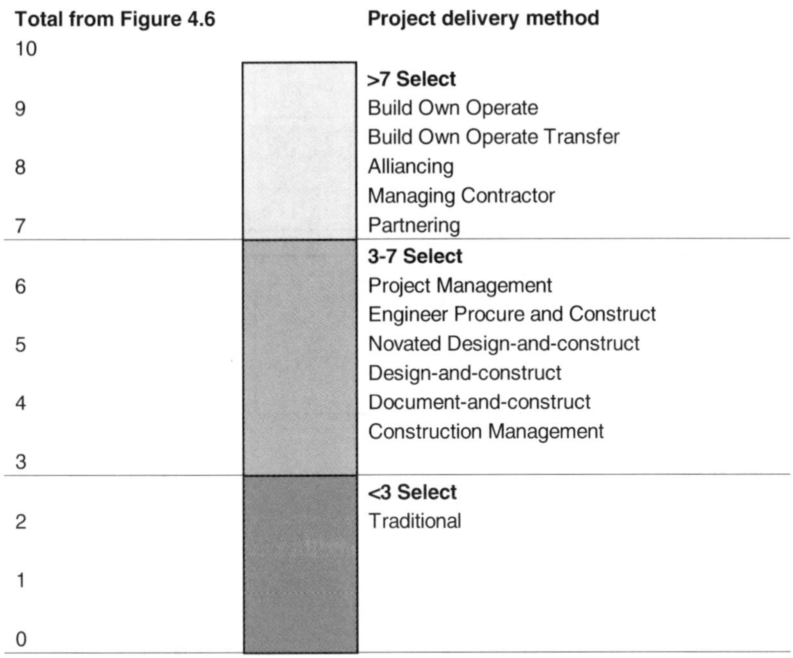

Total from Figure 4.6	Project delivery method
10	
9	**>7 Select**
	Build Own Operate
	Build Own Operate Transfer
8	Alliancing
	Managing Contractor
7	Partnering
6	**3-7 Select**
	Project Management
	Engineer Procure and Construct
5	Novated Design-and-construct
	Design-and-construct
4	Document-and-construct
	Construction Management
3	
2	**<3 Select**
	Traditional
1	
0	

Figure 4.8. Selection of delivery method (after ACA, 1999).

To use Figure 4.10, those issues that are considered critical (carry the largest weighting for that project) to the owner are identified, and then the delivery method options are examined against these issues. In all cases there is a need to consider the risk analysis further for the detail within individual contracts.

4.2.1 CASE STUDY – CONTAMINATED SITES

Introduction

The work practices used in the past (and some still in use today) have created a large number of contaminated sites. The soil and/or groundwater on such sites are contaminated with one or many different pollutants.

Several factors, including:
• increased community concern for the environment,
• human health concerns,
• liability for contamination,
• increasing demand for old industrial land for residential use,
have allowed the development of an industry to assess the contamination status of properties and remediate those that are contaminated.

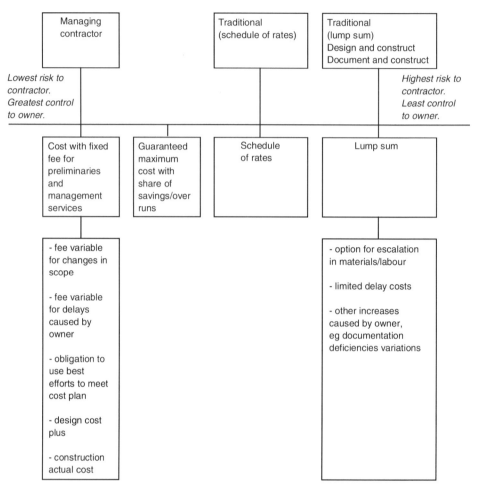

Figure 4.9a. Out-turn cost (final cost to owner) risk allocation – delivery method, payment type and contractor responsibilities (Department of Defence, 1992).

The industry has three major players:
1. Environmental consultants who perform site assessments.
2. Remediation contractors who 'clean-up' sites.
3. Contaminated land auditors who 'sign-off' on clean sites and ultimately take liability.
Historically, auditors and consultants usually work for engineering or specific environmental consultancies, whilst remediation contractors are often large earthwork or construction contractors, although there are a number of specific remediation contractors now entering the market with high technology solutions.

Phases in site assessment/remediation

The phases in site remediation are commonly:

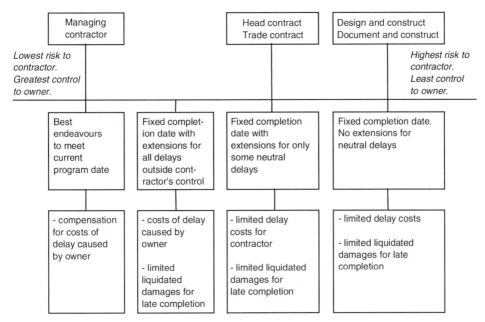

Figure 4.9b. Time risk allocation – delivery method, payment type and contractor responsibilities (Department of Defence, 1992).

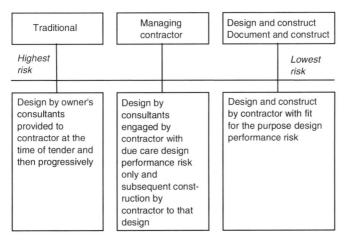

Figure 4.9c. Design risk allocation – delivery method, payment type and contractor responsibilities (Department of Defence, 1992).

Issue	Issues regarded as critical	Tradi-tional	Design development and construction	Design novate and construction	Design and construction
Time impact of tender period		L	M	M/H	H
Cost and/or time impact of individual design changes		L	M	M	H
Impact of design changes on original contract price		L	M	M	H
Cost and time impact of coordination difficulties during design/documentation		H	M	M	L
Overall cost impact of variations due to documentation errors		H	M/L	L	L
Potential for lesser design quality		L	M	L/M	H
High level of contractor's contingencies which add to tender price		L	L/M	M	M/H
Cost impacts due to latent conditions		M	H	H	L
Potential for large time overruns		M	M	M	M
Inability to fast-track		H	M	M	L
Large number of contractor's claims		H	M	M/L	M/L

Figure 4.10. Choosing a delivery method (after New South Wales Public Works Department, 1993). H = high; M = medium; L = low.

- Desktop investigation of site history to determine the operation that could have contaminated the site (Stage 1).
- Soil and/or groundwater sampling and analysis to determine the contamination status of the site (Stage 2).
- Preparation of a remediation action plan (RAP) to make the site safe for its intended land use.
- Remediation of the site.
- Confirmation that the site has been adequately remediated (Validation).
- Independent audit of the site to assess the completeness of investigations and remediation, and verify that the land is suitable for its proposed land use.

During the process, all the work can come under the scrutiny of the government environment regulatory bodies, and/or the local council. In relation to property transactions, investors and banks of the potential purchasers also take a keen interest in the assessment and/or remediation of the site.

Delivery methods used for remediation

Specialised knowledge is required to understand the problems posed by contaminated sites, especially when dealing with regulatory bodies. Many organisations do not have the expertise to communicate effectively with the regulatory bodies regarding site contamination issues, or manage a clean-up (with the exception of some property developers and larger industrial companies). Therefore they must employ consultants.

Table 4.3. Common delivery methods based on size and complexity of remediation project.

Size	Complexity	Procurement method	Comments
Small	Low	Design-and-construct	Consultant investigates site, prepares RAP and performs or subcontracts remediation
Large	Low	Traditional (Sometimes project management)	Consultant investigates, 'designs' RAP. Owner then contracts-out remediation work, with ongoing technical monitoring by consultant. Sometimes project management delivery is used where consultant also acts as project manager
Large	High	Design-and-construct	Consultant investigates and decides what has to be remediated (concept design). This is tendered out to contractors who must produce a detailed remediation strategy (which may involve detailed technology design)

The method of delivery varies, depending on the project size and complexity. These are listed in Table 4.3.

Regardless of the method used, project size or project complexity, an unusual relationship exists between the contractor and the auditor. The auditor is a consultant to the owner, who checks the work of contractors (and other consultants) and notes deficiencies. They do not provide any design functions or management functions to the owner, purely provide advice.

Important factors in selecting a method

The choice of an appropriate delivery method is based on the management of risk. There are considerable technical and financial risks associated with any remediation projects, particularly involving the accuracy of site investigations and the effectiveness of remediation. A real difficulty is in establishing volumes of contaminated soil.

Therefore owners choose a method to minimise their risk.

Why use the traditional delivery method?

The traditional method, where design (i.e. the RAP) is completed by the owner or its consultant is best suited to projects where the remediation process is not highly technical. This is usually the case in the situation involving mainly low level soil contamination – where the soil is just dug up and transported to a landfill.

The following point to using the traditional delivery method:
- The optimum remediation strategy does not require input of the contractor.
- The consultant is used by the owner to provide impartial advice and monitoring.
- The owner wishes to manage the interface between the design (i.e. consultant) and the construction (i.e. remediation contractor).
- The RAP is substantially completed before remediation commences.

Against using the traditional delivery method, there is usually potential for variations because when one area is dug, more contamination may be evident. Provisions for managing this need to be made in contract negotiations.

On smaller sites, it is likely that the owner will give the whole job to a consultant to manage (using the design-and-construct method). However on larger jobs, where a significant amount of work would have been performed by, for example, an earthworks subcontractor (subcontracting to the consultant), the owner can eliminate the management fees (usually placed on the subcontractor) by directly contracting (in this example) to the earthworks contractor.

The traditional delivery method also seems to be used by many owners specifically to reduce the chance of obtaining biased advice from their consultants. There may be a perception that if the consultant is also completing the remediation, it may overestimate the remediation required in order to maximise its return.

Complicated remediation using design-and-construct

The design-and-construct method is an ideal method for dealing with complicated remediation projects.

Site assessments are performed by owners (through consultants) to determine:
* A concept design (approximate volumes to be cleaned up and suitable techniques).
* Performance criteria (i.e. clean up levels).

A contractor is then used to design a method to clean up the site to the required levels using appropriate techniques.

The real advantage to the owner is that it passes most of the risk to the contractor. Whilst variations may be allowed to the contractor based on additional problems encountered in the soil and/or groundwater, the risk of cleaning up known problems to a specified level is passed to the contractor. This applies particularly to highly technical groundwater remediation projects and soil projects where disposal to landfill is not permitted.

Project management method

The project management method is also used on remediation projects. It is useful where the owner does not have the time or expertise to manage its contractor effectively. A practice is developing whereby environmental consultants are doubling as project managers. There is a danger that the consultants (scientific experts) may lack the necessary project management skills. But if they do have these skills they can provide a very economical solution for the owner – supervising the contractor whilst providing technical advice.

The likelihood of misunderstanding is also reduced, due to the project manager and the contractor speaking the same language (especially where an owner is inexperienced). The downside of this is the potential conflict between technical experts who have different approaches.

The trend toward design-and-construct

The trend appears to be moving towards the use of design-and-construct contracts for remediation. As both the 'do nothing' option and the relatively easy 'dig and dump' remediation are becoming less acceptable to the community and regulators, new innovative remediation techniques and technologies are being sought.

The rationale behind this is that owners only have a limited knowledge of the remediation process, whilst many firms are striving to develop unique solutions to gain competitive advantages. Therefore it is logical that the remediation not be constrained by a design imposed by the owner or the owner's consultant, which ultimately could be less successful and more costly than if design was passed to experts. Also, with more involvement of the contractor in the design process, less risk is borne by the land owner.

The advantages espoused for design-and-construct contracting are:
- Completion in shorter time frames.
- Responsibility shifting.
- Construction delays limited.
- Design changes readily accommodated.
- Time spent procuring services reduced.
- Design can be developed around specific equipment components.
- Process has inherently less conflict (as only one firm).

Many projects involve remediation trials that may be best achieved by the same firm scaling up the trials to full remediation works.

EXERCISE
Evidence is pointing towards design-and-construct delivery as optimal for remediation projects, especially when they are complex and require advanced technological solutions. However it must be borne in mind that all remediation projects are very different and often contractors face uncertain and changing on-site conditions. As a result, both contractors and owners need to be flexible in the approach they take and must keep the end goal in mind – the clean up of the site to acceptable levels. How do you incorporate flexibility into the design-and-construct arrangement?

Reading

Manuele, V.O. 1995. Avoid Dangerous Surprises During Site Cleanups. *Chemical Engineering*, pp 125-128.

Ruff, C.M., Dzombak, D.A. & Hendrickson, C.T. 1996. Owner-Contractor Relationships on Contaminated Site Remediation Projects. *Journal of Construction Engineering and Management* 122(4): 348-353.

Tunnicliffe, P., Swatek, M. & McNeice, T. 1995. Design/Build Benefits Bioremediation Projects. *Pollution Engineering*, p 32.

4.3 TRADITIONAL

The traditional ('construct only') delivery method is perhaps the most common form of delivery method used.

Under the traditional method, the owner uses in-house staff or engages consultants to prepare the concept design, the detail design and the contract documentation for the work. The owner then enters into a contract with a general contractor to construct the work. During the construction, consultants may be engaged to provide advice to the owner, to carry out an inspection, monitoring and control role, and to act as certifiers.

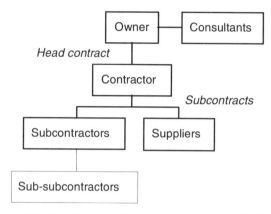

Figure 4.11. Traditional method (showing contractual links).

The contract between the owner and the contractor is referred to as a 'head contract' or 'main contract', while that between the contractor and subcontractors ('specialist contractors' or 'trade contractors') as 'subcontracts'. The consultants (if outsourced) have contracts for services with the owner.

Subcontracting

The contractor divides the work up, typically into trade groupings, and subcontracts the smaller work portions to subcontractors. Subcontractors may in turn divide their work up into still smaller portions and sub-subcontract this out. (Note, the term 'subcontractor' may be used by some people to describe both the real subcontractors, and the sub-subcontractors and sub-sub-subcontractors.) The proportion of work done by the contractor to the work done by the subcontractors, on any given project, varies from close to 0% to close to 100%, depending on the available skills and resources of the contractor. In the former case, the contractor is adopting a managerial role only; in the latter case, the contractor is using its own forces. Whether to use or not use subcontractors is the prerogative of the contractor, but may be influenced by owner requirements. Historically there has been a trend away from contractors using their own forces to using subcontractors, who with time have become more and more specialised.

Subcontractors have a legal relationship with the contractor, who in turn is responsible to the owner for the performance of the subcontractors.

The contractor likens itself to the 'meat in the sandwich' between owner and subcontractors. An alternative view might say however, with respect to some contractor behaviour towards the owner and subcontractors, that the contractor is the person 'trying to put meat on its sandwich'.

Combined consultant(s) and contractor

Sometimes the consultant(s) and contractor have a business relationship or belong to the one organisation.

The architectural and engineering societies oppose this mode of construction, principally because such contracts are allegedly contrary to professional practice. They feel that in allying his interests with those of the contractor, the architect-engineer is removed from his traditional position as an agent of the owner.

(Clough, 1960)

Payment

Such a contract may be on the basis of a lump sum, provisional bill of quantities or schedule of rates, or more rarely on a prime cost basis. (Note, though, that some people loosely use the terms 'lump sum', 'hard dollar' and 'traditional' interchangeably.) Subcontracts would commonly be lump sum. Design services may be on the basis of a negotiated fixed fee.

Usage

The traditional method may be the optimum method for projects where the following requirements are satisfied or substantially satisfied:
- Where the optimum design for the project can be established without involving the prospective contractor or specialist subcontractor.
- Where the detailed design is completed in advance of tendering and the allocation of construction funding.
- Where the owner prefers to manage the interface between the detailed design/documentation and construction, and (where applicable) to select and engage the consultants and have them directly responsible to the owner.
- Where the owner requires consultants to provide advice and monitoring of the project through the design, documentation and construction phases.
- Where the time available for the project is such that the detailed design is complete or may be substantially completed before construction commences.
- Few variations to the design are anticipated to be required during construction.

It is recognised that certain owners use the traditional method for projects that do not satisfy or substantially satisfy some of the above dot points. However, the greater the departure from these dot points, the greater the likelihood of project cost and time increases, claims and disputes.

The traditional method may not be appropriate for some forms of fast-track projects because the traditional 'arms length' relationship between the design team and the construction team is inappropriate when the design is carried out concurrently with construction.

It is essential that the design team has broad site experience covering constructability knowledge, material and manpower availability, industrial relations and safety aspects. If this is not available it can be achieved by appointing a contractor as a consultant on a fee basis:
- As a construction consultant, in a similar relationship to the owner as the design consultants and other specialist consultants.
- The consultant to assess and advise on matters such as constructability, reduction of

problems in design and documentation, costs, program, contract package, selection of a contractor etc.   (based on NPWC/NBCC, 1990)

Advantages – owner's perspective

- With complete documentation, and a lump sum basis, the final cost is known to the owner, before work begins; tendered prices should be realistic; possibly the best, of all the delivery methods, in being able to forecast the project end cost, though not necessarily the lowest cost.
- Design changes are restricted.
- The contractor assumes most risk associated with the construction.
- Full documentation should ensure expected quality outcomes; possibly better quality outcomes than other delivery methods.
- Contractors can be selected competitively.
- Possibly the lowest overall risk, of all the delivery methods, to the owner (though this is difficult to establish on anything other than particular projects).
- The traditional delivery method is well known, accepted and supported by many years of experience.

Disadvantages – owner's perspective

- Construction cannot start until design and documentation are complete; long lead time; this leads to an extended project duration and possible financial risks; fast-tracking not possible. (Most other delivery methods allow fast-tracking.)
- The contractor's skills and knowledge are not used in developing the design – arms length relationship of designers and constructors; constructability or buildability issues may arise; developing the design independently of construction is suboptimal; possibly increased costs and time. (However it is possible to have construction knowledge input at the design stage.)
- Little control over cash flow.
- The contractor is able to pass on risk to lower level subcontractors.
- Possible misunderstandings between contractor, and suppliers and subcontractors, over prices, workmanship and extent of work, particularly with last minute quotations before the contractor submits its tender.
- Requirement for complete documentation opens up possibilities for errors and omissions, and hence potential for claims; complexity of documentation opens up potential for claims: *the capacity to pass on the risk of design and deficiencies in design documentation is limited to the design consultant's failure to exercise reasonable care and skill, whereas the risk of design assumed by a contractor undertaking design in other standard forms is the higher and more onerous risk of warranty of fitness of the design for its purpose.*
- *No single line of responsibility for design and construction, thus the [owner] has the task of sorting out whether a defect is attributable to design or construction*; designer and contractor may be in adversarial roles.
- *Commitment of resources by the [owner] to manage and co-ordinate the interface between design and construction.*   (Department of Defence, 1992)

4.3.1 CASE STUDY – HOUSE CONSTRUCTION

The builder was invited to tender on proposed alterations to a terrace house by its owner. The owner had completed the design of the extension and architectural plans had been submitted to council for approval. The builder annotated these plans indicating the extent of the work, inclusions and exclusions.

A price for the work was agreed and a contract signed. The builder was to provide a fixed cost for the extensions to lock-up. All walls, floors and ceilings were to be lined and electrical works completed. The items of fit-out were presented in a schedule of rates as prime cost (PC) items. Work commenced in August 199X.

The timber floor to the rear of the house was found to be rotted and its replacement was agreed to as a variation. No subfloor access was available prior to demolition of the rear wall/floor. The builder noted this and the costs in correspondence to the owner. A signed and dated copy of the variation was returned to the builder prior to work commencing on this area.

As the end of the year approached, the variations continued, and the PC items were still not resolved. The builder proceeded with all agreed work and continued to update the program. He made diary notes regarding discussions with the owner.

At the time work ceased in February 199Y, the builder had completed the scope of work set down in the original contract, as well as $18,500 worth of variations. The owner, believed that the liquidated damages clause of the contract should apply from the period commencing in December as this was the original program date for practical completion. She also disputed the builder's payment for the new timber floor. The argument raised was that the builder should have noted this work in his original inspection. As such, this work should have been included in his tender. They agreed that an adjudicator was required.

The adjudicator's findings may be summarised as follows:
- The cost of the new floor was not present in the agreed scope of work.
- It was not reasonable for the builder to absorb the cost of this work.
- The owner was paying no more than she would always have had to pay.
- The owner had agreed to this variation at the time in writing. The builder was due for full payment of agreed variations.
- The period of time to complete the original scope of work was slightly less than the original program period. The inclusion of variations had extended the date of practical completion. Again, this was at the owner's request and written agreement. Liquidated damages did not apply.

EXERCISE
1. The builder and the owner had agreed upon a lump sum contract for the additions to lock-up stage, using the traditional delivery method. No schedule of rates, or quantities were provided. Fit-out items were included in the scope of works, but presented as PC items. The approach to procurement also included tendering to maintain competition. Some reasons for adopting this approach included:
   - It is a standard approach for this type or small residential work.
   - The owner's costs for producing contract documents is minimal.
   - 'Off-the-shelf' contract documents exist.
   - The cost involved, to the owner, in producing a bill of quantities is avoided.

- The time involved in tendering was relatively short due to the straightforward nature of the job, its structure and contracts.
- The owner was able to use the expertise and buying power of the builder for items of fit-out.
- Placing the fit-out items as prime cost items gave the owner additional time and should have reduced the risk of budget overrun, as items could be reduced or omitted if required.
- The owner was able to assess the workmanship as work proceeded, and negotiate variations with the builder already established on site, thus saving time and re-establishment costs.
- For the builder, the only competition was during the tender process and this was limited.
- Negotiation and decision making was efficient.

Suggest other reasons.

2. Some negatives with the approach included:
- The builder was managed by a person with little or no experience in construction. The owner's inexperience led to false expectations of time and cost.
- Extended program times are always a source of potential dispute.
- The only fixed price was for the basic structure. Prime cost items could increase, as well as decrease, and this is a risk to the owner if not properly managed.
- The builder's other projects suffered delay due to the owner not choosing PC items until the last opportunity, and involving the builder in disputes.
- Regardless of who was right or wrong in the dispute, the builder lost future work from the owner and people influenced by her.

Suggest other disadvantages.

4.3.2 CASE STUDY – ROAD CONSTRUCTION

Historically, a road construction authority used only the traditional delivery method in all its construction projects. Detail drawings and specifications were prepared by experienced in-house staff. The construction was done by contractors and managed by in-house staff.

In recent times, external consultants are involved in preparing the detail drawings and contract documents. The construction contracts are generally managed by in-house staff. However some contracts are managed by external management consultants.

Due to a greater demand for fast-tracking of projects and a shortage of in-house expertise, the authority is looking to move towards design-and-construct, using performance-oriented contracts, to minimise maintenance during the life of the roads.

However at the present time, the majority of roads are built under the traditional method, while parts are constructed under design-and-construct. The selection of the preferred delivery method is based on 'best value for money'.

Under the traditional method, the concept design and the environmental study are carried out by in-house staff, with some external assistance. The detail design and contract documentation are prepared by external design consultants under the authority's coordination. Tenders are called from pre-qualified contractors for major contracts, and public tendering is used for minor contracts. Generally the minor contracts are valued less than one million dollars. For both major and minor contracts, the authority allows alternative tenders.

EXERCISE
How might 'best value for money' be interpreted in the context of delivery methods?

4.4 DESIGN-AND-CONSTRUCTION

Design-and-construct(ion) (D&C) may be carried out with different degrees of involvement of the owner, and under different titles.

- Where the owner develops the design to a significant extent and then passes it over to a contractor to complete the detail, this may be referred to as *detail design-and-construct(ion)*.
- *Design-develop(ment)-and-construct(ion),design-document-and-construct(ion)*and *document-and-construct(ion)*, but using terminology that gives added emphasis to certain activities. The proportion of design that the owner does is flexible; the owner may carry out conceptual design only, or something more.
- Where the contractor does essentially all the design based on a brief from the owner, this may be referred to as *design-and-construct(ion)*. It may also be called a *performance contract* where the owner stipulates the performance of the end-product and the contractor provides an end-product to satisfy this.
- *Turn-key* arrangements are similar to design-and-construct. The whole design-and-construction package is outsourced and the facility or product presented in a ready-to-use form (complete except for the ribbon-cutting ceremony – 'turn-key ready') for the owner. A *package deal* is provided to the owner. Turn-key is different to design-and-construction if commissioning and handover are not embraced within the term 'construction'. Some people make the distinction: if commissioning and handover are present, it is turn-key; if they are not present, it is design-and-construct. But this distinction is not universally practised. The terms design-and-construct and turn-key are interchangeable in general usage. An extended version of the turn-key arrangement may include, at the front end of a project, some involvement of the contractor in such matters as feasibility studies, land acquisition, financing etc.
- The term *design-manage* might be used where the organisation responsible for design and construction employs consultants to do the design, and independent contractors to do the construction, while preserving a largely managerial role for itself.
- The *managing contractor* method involves the contractor managing the design consultants, subcontractors and preliminaries, and having costs reimbursed, as well as being paid a (usually fixed or a percentage) fee to cover overheads and profit. The level of responsibility expected of the contractor and the level of risk taken by the contractor are less than in full design-and-construct, with the intent that the contractor should act more in the owner's interests. It is akin to design-and-construct on a cost-plus payment basis. All the construction is done by the subcontractors, and none by the contractor.

Commonly, the design and construction phases are overlapped in an attempt to reduce the total project time. A phased program for construction is used.

Design-and-construct contracts are commonly used in developing infrastructure for the process industries, in heavy industrial projects, and significant building development work.

D&C is popular with investors and developers because it can deliver a product within predetermined cost and time constraints, and hence gives an assured return on investment. Life cycle and design-variety issues may be sacrificed in the process however. D&C is one-stop shopping.

Payment

Commonly, but not exclusively, a lump sum payment to the contractor is used. Provision for rise and fall in prices can be included. Guaranteed maximum price (GMP) also has some popularity. Schedule of rates does not lend itself to design-and-construct contracts. (The managing contractor version uses cost-plus payment arrangements, perhaps based on a target cost with incentives for quality, under-budget and before-time delivery.)

While a bill of quantities is not prepared, changes to the scope of work (variations) may be priced on a schedule of rates.

General

Where the design is left entirely to the contractor, the owner avoids the risk associated with design errors and the flow-on project price increasing over the tendered lump sum. Owner-caused delays can be reduced through minimising owner involvement, and relying on the contractor to produce the facility, to the owner's goals, for the fixed price.

Instalment payments to the contractor can be linked to milestones, rather than on work done. This better defines the owner's budget commitments over the duration of the contract. The D&C contract tends not to lend itself to the conventional method of payment based on work done, because it is difficult for the owner to establish the real cost of work done. Using milestones, the contractor directs its resources to completing the milestones, compared with payment for work done where the contractor does the most profitable work first to improve its cash flow.

With such brief documentation available to the contractor at the time of tendering, it is difficult for the contractor to estimate the cost of the work exactly, and this necessitates the inclusion of contingencies in the bid. To balance this extra cost, the owner hopes to obtain a completed facility sooner, and thereby possibly get an earlier return on its investment.

There is the potential for the contractor to compromise on quality should the contractor perceive not making a sufficient return. This may necessitate the owner or a consultant taking on an inspecting and supervising role.

On a lump sum basis, the risk associated with the completion cost and time is essentially taken by the contractor, as in the traditional delivery method. This risk can be reduced if rise and fall and variations are allowed. As well, there is the risk associated with design problems.

Conditions of contract

Standard contract conditions exist for design-and-construct work. In addition, conditions of contract used for traditional delivery may be used for design-and-construct delivery, with slight modifications. It is argued that, no matter what the method of delivery, the contractor has to perform some design, even if it is only temporary works or guessing a detail that the designer has not cared to document because it is not overly important. Different delivery methods differ in the quantity of design that the contractor performs. Providing that the conditions of contract refer to the term 'work', then both design and construction can be included within this term; it remains for some other contract document to spell out what 'work' means and its extent.

Modifications of conditions of contract for traditional delivery to suit design-and-construct delivery could include, for example, deletion of clauses (in the traditional delivery contract) relating to latent site conditions and temporary works, certain grounds for extension of time and variations, and design issues (e.g. approvals from government authorities, mistakes, copyright, ...) being the responsibility of the owner.

Some people, however, argue that conditions of contract need to be specially developed for the design-and-construct case, and that modifying conditions of contract intended for traditional delivery is unwise.

The design-and-construct contract needs to make reference to the contractor's responsibility that the completed facility will be 'fit for the purpose intended'. With the contractor just satisfying the contract documents alone, without reference to 'suitability for purpose', this may give the owner an unusable facility, and lead to a dispute between the owner and contractor. The owner relies on the expertise of the contractor to produce a facility suitable for its intended purpose. With D&C, the owner obtains a suitability-for-purpose warranty from the contractor.

In a design-and-construct contract there is always the problem that if the [owner] reserves the right ... to approve the contractor's drawings, approval, refusal to approve, failure to disapprove, or delay in approving may result in a variation, a claim for delay costs, a waiver or another claim. The answer is not to require drawings to be produced for approval but simply to require that not less than a certain time before the contractor intends to use any drawing, the contractor must give a copy to the [owner]. The [owner] will then have an opportunity to order a variation or to object that the drawing is not in accordance with the contract. It is important to say that the [owner] is not required to inspect the drawings or to notify the contractor of any error or departure from the requirements of the contract and that failure of the [owner] to comment on the drawings or any particular item in them must not be taken as a waiver of any requirement of the contract. The [owner] must be careful not to indicate expressly or by implication from conduct that a drawing is suitable for use.

What is or is not a variation often causes arguments in design-and-construct contracts. There is a risk that in approving of a design suggestion from the contractor, the [owner] or [owner's representative] may be taken to have approved of a variation.

(Uher & Davenport, 1998)

Express exclusion of such possible variations, or express statement of what only constitutes variations may be necessary in the conditions of contract.

Owner's brief

The owner's brief is one of the important contract documents. It is additional to the conditions of contract, specification and any drawings. It sets out the owner's requirements. It articulates the purpose for the end-product or facility, and if this articulation is done clearly, transfers liability for the end-product to the contractor. It may be prepared by the owner or the owner's consultant, and in some cases by the contractor itself.

The brief may contain:
• Performance criteria.
• Functional requirements (the term 'functional brief' is sometimes used).
• General requirements (mandatory and indicative).

Having outlined its requirements, the owner then relies on the contractor's skill and judgement to give the desired end-product.

There is always the concern as to how much detail is necessary in the brief. The more complicated the end-product, the more detailed and comprehensive it would seem to have to be. Completely defining particular elements however, removes the contractor's liability for the elements meeting the owner's purpose.

From the owner's viewpoint, more detail gives:
• Less opportunity for variations.
• More certainty regarding the end-product.
• Less chance of under-design.

From the contractor's viewpoint, less detail gives:
• More flexibility in its approach, and containing costs.
• Better opportunity for constructability input.
• Less conflict with the owner.

Where the owner's brief is inadequate, the design or work do not meet the owner's expectations, and extra work is required, the contract sum would be adjusted. The owner bears the risk associated with an inadequate brief.

Failure to ensure that the owner's brief is stated in clear, objective, performance terms may cause disputes as to whether the contractor's design satisfies the requirements of the concept design and performance specifications provided by the owner. Different interpretations are possible between owner and contractor, even in the most carefully prepared briefs; this can result in friction between the owner and the contractor.

Failure to properly identify the end-product requirements in the tender documents may lead to the 'loading' of tender prices by tenderers.

At tender time, the owner selects, not only on price and the tenderer's qualifications, but also the way the tenderer addresses the owner's brief. For design-and-construct this will involve developing a conceptual design.

Variations

Variations can become a source of conflict in D&C contracts. The interpretation of what constitutes a variation might be wide, and hence needs clear definition in the conditions of contract.

Design changes may be brought about through:
• Changes to the owner's brief or the contractor's design by the owner.
• Changes suggested by the contractor.

Changes requested by the owner may adversely impact on the contractor's design responsibilities, and the contractor's construction method. Changes initiated by the owner could be expected to be costly where they impact on the contractor's program. Changes, whether requested by the contractor or the owner, may be difficult for the owner to cost. Design changes could be expected to impact on the construction, and this impact may be difficult to cost. Where the total cost reduces because of a contractor-initiated design change, consideration might be given to sharing the cost savings between contractor and owner.

4.4.1 DETAIL DESIGN-AND-CONSTRUCTION

With detail design-and-construction, the owner uses in-house staff or engages design consultants to prepare a concept design and performance specification. The owner then enters into a contract with a contractor/designer to prepare the detail design and documentation in accordance with the concept design to satisfy the performance specification, and to carry out the construction and commissioning. The contract is usually for a lump sum or guaranteed maximum price that may be subject to adjustment for various neutral risk factors (Fig. 4.12).

Usage

Detail design-and-construction may be the optimum method when the following requirements are satisfied:
- The owner wishes to develop the concept design as well as the performance specification.
- The owner requires the contractor to be responsible for the detail design and documentation, and construction and commissioning.
- The owner has well-established standards for details and finishes, enabling a thorough specification; the design brief is clear.

Advantages to the owner

- The owner avoids some coordination responsibilities, such as occur in the traditional method; the owner transfers coordination responsibility and risks to the contractor.

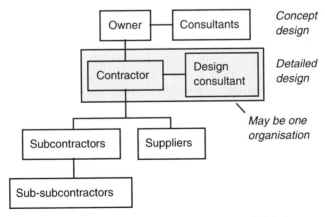

Figure 4.12. Detail design-and-construct method (showing contractual links) – contractor as the lead player. The alternative where the design consultant is the lead player, rather than the contractor, is also possible.

- The owner retains control over the concept design; the owner transfers detail design responsibility to the contractor.
- Reduced (over the traditional method) scope for extension of time and variation claims.
- Specialist and proprietary designs, available in the marketplace, become available to the owner; scope for innovation broadened.
- Reduced (over the traditional method) owner resources required.
- Potential savings (cost and time) through the contractor coordinating the design and construction processes and resources, tailoring design to construction, and a faster overall project.

Disadvantages to the owner

- The cost to the contractor of tendering is high, reducing the size of the tendering pool; as well, the pool of competent contractors may be further reduced over a construct only delivery method.
- Potential for ambiguity of interpretation in the owner's brief; potential for lower quality levels (over the traditional method); potential for extra cost to upgrade the quality to that expected; potential for disputes over 'fitness for purpose' of contractor's solution.
- The tender price could be expected to be 'loaded' to cater for the extra risks carried by the contractor, and the extra cost of tendering.
- Changes requested by the owner can be costly, because of the interference to the contractor's possibly fast-tracked program and accepted contractor's design.

General

The delivery method avoids the contractor having to do investigatory and outside authority-approval tasks (when compared with full design-and-construct).

In a detail design-and-construction contract, construction can start, at the contractor's risk, prior to the finalisation of the detail design, thereby reducing the project time to a minimum. This concurrence of detail design, documentation and construction is possible because the design team and the construction team can work closely together and avoid the formality necessary between the design team and the construction team when design is carried out at 'arms length' to construction.

In a detail design-and-construction contract, the contractor is usually not entitled to an extension of time or increase in the contract sum for variations to the detail design caused by the progressive development of the design during the construction phase, or for late supply of design information, lack of coordination between documents and errors in the bills of quantities, except where these are caused by the owner directing variations to the specified quality and performance requirements. There is an incentive for the detail design and documentation to be fine-tuned by the use of constructability studies (value management/analysis/engineering) to ensure that the adopted detail design-and-construction methodology minimises time and costs, whilst complying with the owner's specified requirements.

In detail design-and-construction, the control of the design passes from the owner to the contractor. Care should be taken to ensure that the concept design and performance specifications prepared by the owner are stated in clear, objective performance terms, because failure to do this may cause disputes as to whether the contractor's detail design satisfies the requirements of the 'concept design and performance specifications'.

4.4.2 DESIGN-AND-CONSTRUCTION

In this delivery method the owner contracts directly with an organisation which is responsible for providing the design, documentation, construction and commissioning for a project to satisfy the owner's specified performance and quality requirements (Fig. 4.13).

Some owners see D&C as a way of divesting all project obligations to another party, but allocating blanket risk may cause disagreements. Some owners see D&C as a way of making up for lack of experience in, or lack of resources for, undertaking projects.

With design and investigation work necessary for tendering, the cost of tendering can be high. This may be partly offset by only having selective tendering from a small number of prequalified tenderers, or the owner reimbursing in whole or in part the cost of tendering to the tenderers. The latter option appears to occur only rarely. The indefiniteness connected with having only partial design and investigation available at the time of tendering, could be expected to be reflected in the tenderers' prices. Having each tenderer carry out design and investigation separately results in much duplicated effort and cost; an investigation might be arranged by the owner for common release to all tenderers.

Payment

Commonly the design-and-construct work is done for a lump sum or guaranteed maximum price which may be subject to adjustment for changes initiated by the owner, and may be subject to adjustment for various neutral risk factors.

The tender evaluation would not only consider price and the contractor's capabilities, but also the proposed design and its technical merits.

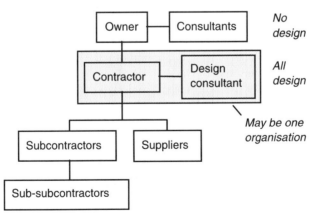

Figure 4.13. Design-and-construct method (showing contractual links). The alternative where the design consultant is the lead player rather than the contractor is also possible.

Usage

Design-and-construction delivery may be most suitable for projects that have:
• Simple and concise briefs.
• Few complex requirements.
• Little likelihood of changes after contract award.
But is also used for large and very complicated projects.
Usage possibly means:
• The owner's brief for the project can be properly identified and expressed in objective, performance terms; the owner's end-product requirements can be properly identified at the time of calling tenders.
• The owner has well-established standards for components, details and finishes.
• The owner does not have available in-house design resources; the owner requires tenderers to develop concept design(s) to satisfy the owner's performance brief, and then the selected contractor to develop the design detail. Tenderers may be encouraged to develop alternative concept designs, for cost saving, aesthetic or other reasons.
Construction can commence, at the contractor's risk, prior to the finalisation of the detailed design, thereby reducing project time to a minimum.

It may be difficult to get a contractor to commit to a project completion date if matters such as approvals from authorities are included in the contractor's scope of work. The time involved in the approval process can be variable depending on the efficiency of the authorities involved.

The control of design quality can be difficult.   (based on NPWC/NBCC, 1990)

Advantages over traditional method

Arguments in favour of design-and-construction type methods over traditional construction only include:
• Possible reduced total time of design and construction by integrating these stages and possibly fast-tracking the process.
• Removal of the separation of the design function and the construction function; the design can be better tailored to construction practices and equipment, to suit availabilities of materials and work skills. Value management/analysis/engineering (constructability, buildability) should ensure a total design-construction optimisation instead of subsystem design and subsystem construction optimisation. There is no adversarial relationship between designer and constructor.
• The contractor takes the risk with respect to design and documentation inadequacies; less variations and claims could be expected on these matters; less disputes could be expected on these matters. The risk associated with whether the design is fit for purpose, is transferred to the contractor.
• Documentation and detailing could be expected to be less.
• Single point of responsibility for design and construction; contractor assumes total responsibility; less owner managerial role of interfaces. Risk associated with the design-construction coordination role transferred to the contractor; owner coordination role minimal.

- Possibly lower total price based on competitive tendering for the total design and construction package, through possible innovation; encourages innovation.
- When resources to perform the design are not available in-house; reduced demand on owner resources.
- If a tenderer is unsuccessful, the design effort and costs are thrown away. This may force designs to be less conservative – over conservatism may doom the tender.

Disadvantages compared with traditional method

Arguments against design-and-construction when compared with traditional construction only include:
- Some lack of control and inspection role by the owner, particularly of the design and quality aspects; only minimum standards may be received. Contractor's attention to details and aesthetics may have a lower priority than that wanted by the owner.
- The opportunity for the owner to make changes is diminished; variations can be costly.
- The prices tendered may be higher because of the lack of definition of the work at the tender stage, and the additional risks borne by the contractor.
- Lack of definition of work at the tender stage may lead to contractor claims later on for work to upgrade to a higher standard, or work of unacceptable standard.
- The increased cost of tendering to the contractor is thought to be balanced by a decreased cost to the owner of preparing documentation; some owners may contribute to the cost of tendering.
- The possible reduced total time of design and construction may be balanced by possible extra lead time pre-contract, preparing tender documents, undertaking site investigations etc.
- Difficulty specifying, up-front, standards, details, aesthetics and so on, particularly without being too prescriptive or by using a functional brief; brief may be open to many interpretations some of which may be unanticipated and unacceptable to the owner.
- Potential maintenance problems down the track; this can be partly addressed by including a long defects liability period – the contract then effectively becomes *design-construct-and-maintain*.
- Design effort and costs of unsuccessful tenderers are thrown away; the higher costs of tendering may be countered by having a reduced pool of tenderers, but this reduces competition.
- Design-and-construct contracts, though not a fundamental characteristic, tend to be let for larger projects and larger slices of work. This may eliminate smaller contractors from tendering. Together with the practice of prequalification or invited tenderers, the opportunities for smaller contractors are less again. The pool of tenderers is thereby smaller.
- More difficulty in selecting between different tenders, because no two designs will be the same; cannot compare like with like.
- Subcontractors are asked to bid on incomplete documentation.
- Because of the additional risks placed on the contractor, additional insurance is required; bond rates are higher.

- Because of the potentially faster nature of design-and-construct, owners have to find money faster to pay for the work; this may not suit some owners.
- With lump sum or guaranteed maximum price, profitability might be obtained by the contractor through adjusting quality or design; owner surveillance required.
- With few checks and balances on the contractor's performance, the owner may not be made aware of problems, or of design and construction coordination troubles, or of cost or time overruns, or that the ongoing construction will give what the owner expects.

With design-and-construct, the design effort and costs of unsuccessful tenderers is thrown away. This can be addressed partly by limiting the number of tenderers, or negotiating with a sole tenderer. The cost of design may be borne by the contractor or by the designer or shared between the contractor and designer (where the designer is a separate organisation to the construction contractor). In general, only the smallest amount of design work, that allows a tender to be prepared, is done. The figure of 15-30% is sometimes quoted – that is, bidding is based on design 15-30% complete.

Design-and-construct may be inapplicable in many instances, for example involving land acquisition, community involvement and public comment, unusual or one-off projects, and designs that need testing before building.   (based on NPWC/NBCC, 1990)

Consultant's perspective

Many consultants view D&C with unease and suspicion.

Frequently the lead role in design-and-construction is taken by a contractor who coopts the services of a design consultant. This need not be the case, but is a common scenario. Few contractors seem to maintain in-house design personnel, but rather designers are engaged on a needs basis. The design consultant comes secondary to the contractor and is answerable to the contractor, compared with the traditional delivery method where the designer deals directly with the owner. In some cases, the designer is a subcontractor to the contractor. The importance of the design is perceived by the consultant as being diminished in the eyes of the owner. Design control and influence may be transferred to the contractor. Design consultants who also offer services additional to design, such as project or contract management, have their potential services reduced in extent.

Consultants may fear losing money through doing work for tenders that turn out to be unsuccessful. Typically a design may have to be taken to about 15-30% complete in order that a tender price can be worked out by the contractor. Given, say, a one in five chance of winning a tender, this one successful project has to recoup the losses of the unsuccessful tenders, and this may not be possible. Many design consultants will not consider preparing work for bidding when there are more than, say, three bidders. Ideally the consultant would like some reimbursement for pre-tender design costs, or a share with the contractor in any project profit proportional to the risk each party is taking. Where the designer and constructor are in partnership or in a joint venture, this will be taken care of directly.

Consultants may fear contractors who screw down the design costs in order to make the bid competitive or to improve the contractor's profit. Sufficient time and budget for design needs to be allowed by the contractor. Compromising on design time and budget can be counterproductive in terms of smoothness of the construction and its management. In this regard, design-and-construct may not offer any shortcuts in project time. As well, there can be a mistrust of contractor motives by the designer in that there is a belief that

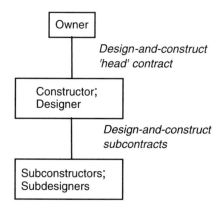

Figure 4.14. Design-and-construct on two levels.

the contractor puts cost before design, with resultant quality implications, particularly long-term.

Both consultant and contractor need to work together as a team for a successful outcome. The design is not separate to construction. The designer and contractor are not adversaries. In this sense, the consultant should not be made a scapegoat by the contractor for things that go wrong.

Design-and-construct on two levels

Design-and-construct may be practised on two (or more) levels. The upper level involving a constructor and designer overseeing subconstructors and subdesigners at a lower level (Fig. 4.14). The designer at the upper level may be novated to the lower level. Third and lower levels are possible. Levels below the top level add more and more detail to the process.

4.4.3 CASE STUDY – HOSPITAL PROJECT

The case study describes the design-and-construct (D&C) contract for the delivery of a major hospital with a total project value of $40M and a construction contract value of $30M. In order to understand the selection of the D&C method, it is necessary to describe the history of the project itself and the effect of this history on the owner's needs and expectations.

The decision to build a new hospital was made by politicians following a tour of the existing old hospital. Usual practice at that time, and the one followed for this project, was the traditional method of project delivery.

Following the appointment of the design team in 199A, the master development control plan was prepared over a six month period. This was three months late in being completed. The subsequent schematic design was completed five months late, and design-development completed six months late. Contract documentation was due for completion in March 199C (being seven months late) but was not achieved because the architectural firm was 'liquidated' in February 199C.

The liquidator took possession of the design documentation and indicated that it would only be given to the owner if outstanding debts were paid. Two months later, while a legal battle to recover the documents was being carried out, an election was announced. Following the re-election of the government, a review of current capital works programs was announced and the subject hospital project was put on hold. The local Member of Parliament reassured the community that the new hospital would be built as promised but would be delayed a couple of years.

In mid-199D, the allocation of funding for the project was announced, but the hospital would be reduced in size from the initially proposed 200 beds to 150 beds. The budget was correspondingly reduced from $80M to $40M. It was also politically imperative that the hospital be completed and operational before the next election.

It was with this political constraint and a desire to try another project delivery method that the owner decided that the hospital would be built under a D&C contract, a first in the hospital arena.

The contract documents and design brief were then prepared over a six week period. The tender documents were issued to a short list of three building companies in August and tenders closed in October 199D. Following an extensive tender evaluation process, the contract was awarded at the end of November to the second lowest tenderer. The accepted price was substantially lower than the pretender estimate.

Consultants (mechanical & electrical engineers, landscape architects, acoustic engineers, building surveyor etc) were engaged to advise the project manager on design and building compliance, progress payment certification and alternative equipment evaluation issues.

In order to demonstrate to the local community that the hospital was on its way, the builder obliged the owner by commencing the bulk earthworks in December. However, as a result of the design being modified over the following months, some the earthworks were inappropriate and rectification works cost the builder a substantial amount of money.

The schematic design was modified in the first month and the detailed documentation including 'user group' review meetings took place over a period of five months. The user group review process was very traumatic for the owner's engineering staff as they were expected to review partially completed documentation in very short periods of time. In some instances, design alterations were made without consulting the user group due to the critical nature of the work.

The builder continued the construction process concurrently with the design process. The structural trades progressed well but problems started to arise with the building services. The proposed heating/ventilation/air conditioning (HVAC) system designed by the consultant was $0.4M more that the builder allowed. The subcontractor designed an alternative cheaper system that was subsequently accepted by the owner. It also became apparent that the services had not been properly coordinated and this was contributing to delays in the construction.

One of the biggest areas of concern for the project was the quality of materials, plant and workmanship. As the construction phase progressed and the true cost became apparent to the builder, the quality aspects were minimised wherever possible. It is worth noting that the architects did not produce a defects list for the project and the builder only rectified those defects identified by the project manager.

The building was due for completion in July 199F but Practical Completion was not given until September. In reality, the builder was still working up until the day before the hospital opened in October.

Twenty months after Practical Completion, the builder was still rectifying defects. Examples of the major defects included re-levelling the operating theatre floors, structural alterations to the steelwork supporting the operating theatre lights, incorrectly built aseptic clean rooms, heat loading problems, acoustic problems, incomplete landscaping works etc.

The above description of the project and its history is only a fraction of the events and issues that occurred over the six years.

EXERCISE
1. How can you address the issues raised, while still maintaining an essentially D&C method?
2. A design 'brief' cannot describe all complexities of a hospital. A builder can interpret the brief so as to carry out the works with a fair degree of flexibility. How does an owner get around this in a D&C contract? How does this affect quality, and life cycle costing, and operation and maintenance issues?
3. What potential causes for delay claims exist in this case?
4. Is the exposure to an owner in a D&C contract different in conventional buildings compared to specialist buildings such as hospitals?
5. Why was D&C delivery preferred over the traditional method for this project?

4.4.4 CASE STUDY – ROAD BUILDING PROJECTS

One road building authority is currently experimenting with the design-and-construct method for its projects. This occurrence has come about through the downsizing of the authority's organisation, in particular the design department.

The authority believes that design-and-construct contracts are a benefit due to the financial savings, but are causing frustration due to the uncertainty of the end-product received.

In early design-and-construct contracts involving the authority, the brief was found to be inadequately detailed with respect to the quality and aesthetics of the end-product. The function and design criteria as detailed in the design-and-construct brief and specification could easily be met, and resulted in many distinctly different end-products.

For example, on one project, the authority wanted a dual carriageway bridge over a creek. A two-span precast beam bridge with a single pier in the creek was anticipated. The contractor supplied a pre-built steel arch over the creek and backfilled. The contractor believed it met the specified design criteria and functionality for the bridge and it was a better solution environmentally, as less disturbance would occur in the creek. The authority had anticipated a more 'aesthetically pleasing' structure similar to surrounding bridges.

The authority felt, at that point in time, design-and-construct contracts were suffering with respect to quality, as contractors would find the cheapest method of construction, which would not always give the result the authority expected, but would nonetheless meet the specified criteria.

In more recent design-and-construct contracts, the authority has spent considerable time refining the brief to reflect more concise criteria for the desired end-product. The brief is now far more detailed in an attempt to clarify exactly the product and standard of quality required.

Although more recent design-and-construct contracts have seen more concise briefs and specifications from the authority, some contractors feel that they are entering into detail design-and-construct contracts, rather than conceptual design-and-construct contracts. As well, recent contracts are limiting the innovation and method of construction in design-and-construct projects.

Ultimately, the owner must receive an end-product that meets all its goals and expectations, regardless of the type of end-product contractors wish to supply.

EXERCISE
1. How can you address the issues raised, while still maintaining an essentially D&C method?
2. How does the owner balance the extent of the design it does versus getting what it wants from a D&C package?
3. In order to address quality, maintenance and ongoing asset management costs, is there any value in going to the design-construct-and-maintain method?

4.4.5 MANAGING CONTRACTOR

The managing contractor method involves the contractor managing the design consultants, subcontractors and preliminaries, and having costs reimbursed, as well as being paid (usually a fixed or a percentage) fee to cover overheads and profit. The level of responsibility expected of the contractor and the level of risk taken by the contractor are less than in full design-and-construct, with the intent that the contractor should act more in the owner's interests. It is akin to design-and-construct on a cost-plus payment basis. All the construction is done by the subcontractors, and none by the contractor.

The managing contractor method borrows a bit from project management and construc-

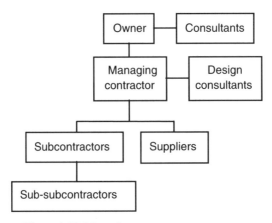

Figure 4.15. Managing contractor arrangement.

tion management methods, as well as from the traditional method. The owner and the contractor share the risks, though most applications seem to involve the owner bearing most of the risk for cost and time overruns, in an attempt to remove any possible adversarial relationship between the contractor and the owner, and in turn hoping the contractor will act in the owner's interests.

The contractor tenders a fee based on a project of broadly defined scope and ceiling cost, though this fee may be varied if the scope of the project varies, or the project is delayed.

The method provides flexibility to the owner to include whatever design aspects or changes it wishes, at any time throughout the project, while using the management expertise of the contractor in planning, control, supervision, administration, coordination and dealing with subcontractors and consultants and industrial relations, as well as technical expertise in product/facility design and construction design.

Some contractor accountability is obtainable if, for example:

- Some standard of care is expected from the design product (the contractor acts like a lead consultant).
- The contractor is responsible for mistakes, or subcontractor claims of its own making.

Otherwise, the contractor is only obliged to doing its 'best endeavours' to complete the project to some target cost and duration.

Subcontractors may be chosen with approval of the owner, or the subcontractors may be nominated subcontractors.

After a number of cases in the 1980s which indicted the greater risk for the [owner] when a subcontractor is nominated by the [owner], the term 'nominated subcontractor' fell out of favour. Parties often strenuously deny that subcontractors chosen by the [owner] or jointly by the [owner] and the managing contractor are nominated subcontractors. Nevertheless, from the point of view of the application of legal principles, the name given to a subcontractor is not determinative of whether the subcontractor is or is not a nominated contractor.

(Uher & Davenport, 1998)

4.5 NOVATION

Novation refers to the substitution of a new obligation for an old one by mutual agreement between the parties. Where there is a contract, a new contract is substituted and this may be between different parties. The consideration for the second contract may involve the discharge of the first contract.

Novation occurs when A, B and C agree that a contract between A and B will be replaced by a contract between B and C, thereby terminating the contract between A and B. Novation is not new. It was known in Roman Law. It is used in construction contracting where the [owner] wishes to enter a contract with a specialist (consultant, manufacturer or contractor) on the basis that when the [owner] lets the main contract, the specialist will become a subcontractor (or consultant) to the main contractor.

(Davenport, 1993)

Most commonly in project work it is practised with respect to design and is termed *design novation*, *novated design-and-construct* or *design, novate and construct* (DN&C). However, it can be used for work other than design, for example manufacture, fabrication or any contractor or consultant service. The following description is in terms of design novation, but generally applies for novation involving any specialist consultant or contractor.

In design novation, the owner engages a design consultant to develop design and documentation to the extent that the owner's needs and intent are clearly identified. Before the design is fully complete, contractors work with the design consultant on aspects of the detail of the design, and tender for the construction and the remainder of the design. The successful contractor, in effect, has a design-and-construction type contract, but using the owner's selected design consultant (Fig. 4.16). The design and construction phases can be fast-tracked. The design consultant effectively takes on a role like a nominated subcontractor.

Novation contracting is a re-birth of ideas associated with *nominated subcontracting*, a term and practice which has fallen from favour.

The traditional method separates the design and construction processes. Novation transforms what starts out as the traditional method to a design-and-construction type method. Figure 4.16 shows the pre-novation and post-novation stages.

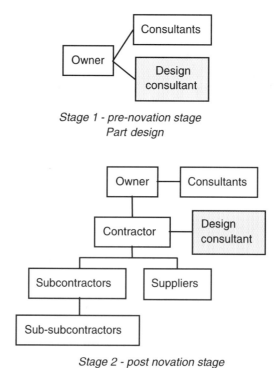

Stage 1 - pre-novation stage
Part design

Stage 2 - post novation stage
Remaining design plus construction

Figure 4.16. Design novation (showing contractual links).

Design in the pre-novation stage might typically be anything in the 30-80% range of the total design. There is associated documentation, sufficient to enable tenders to be called.

In the post-novation stage, the contractor has a direct contractual link with the owner and assumes responsibility for design as in the design-and-construction method. The design consultant's loyalty shifts from the owner to the contractor as a direct reflection of a shift in source of payments. The contractor is obliged to use the original design consultant, and it is this feature which distinguishes the post-novation stage from conventional design-and-construction. In essence, design, novate and construct is similar to detail design-and-construction, but with continuity of the designer throughout the process.

The contractor has some flexibility in the direction the design takes, and can customise the design to suit construction practices, as long as the design stays within the overall contractual requirements laid down by the owner. The owner may have an inspecting role on the design in the post-novation stage but needs to avoid inadvertently accepting responsibility for the design. The owner may retain the right to monitor and comment on the design detail. There may be a tendency for the owner in the post-novation stage to continue to issue instructions to the design consultant.

Payment

Typically such contracts are of the lump sum type, in order to give a reasonably fixed budget for the owner.

Usage

The main use comes for novation contracting where the owner wants to develop the design to a certain level, yet wants single point of responsibility for construction and the detail of the design. It also has applicability where project completion time is a concern, as the approach may be used as a form of fast-tracking. Most of the risk of finishing to a cost and time requirement is transferred to the contractor.

Design novation may find use where:
- The owner wants control over the design direction, yet in-house design resources are not available.
- The design brief can be clearly defined; products and materials can be prescribed.
- Specialist or proprietary designs or construction methods are available in the marketplace.

Advantages, disadvantages

Most of the advantages and disadvantages of novation contracting are the same as design-and-construct, where some design is done before handing over to a contractor.

Design continuity
The owner obtains continuity of design thought throughout the project. Design intent and standards can be maintained. The advantages of design-and-construct are kept while retaining some control over the design.

Design warranty
To the contractor, there is the disadvantage that any design errors made in the pre-novation stage may become the responsibility of the contractor.

Predetermined relationships
The contractor has to live with the designated design consultant and the consultant's deficiencies as a designer. This might be reflected in higher tender prices. How does the contractor deal with inadequate work of the designer? There is a need to match the designer with the contractor. Against this, there could be expected a greater willingness on the part of the owner to accept the design.

Payments
To the design consultant, payments in the post-novation stage come through the contractor; there are associated potential problems of non-payment.

Legal framework
The legal framework has to be thought through carefully to avoid pitfalls. Problems can arise in the parties coming to an agreement over contractual conditions, and how to handle the rights and liabilities of all the parties. Also, what happens in the situation if the owner, contractor or designer become insolvent?

Risk allocation

The risk allocation aspects between the parties are very similar to that in the design-and-construct method where some design is done before handing over to a contractor. Additional risk aspects arise for the contractor because the design consultant is pre-chosen, and for the design consultant because of payments coming through the contractor in the post-novation stage.

Operation

There may be a perception to contractors that design novation represents a significant risk transfer, and this may be reflected in the contractor's price. Until this method becomes reasonably commonplace, this situation may persist and this in turn may operate against people using the method.

4.5.1 CASE STUDY – ONE COMPANY'S EXPERIENCES IN THE BUILDING INDUSTRY

The case study outlines the decisions made, in one company, on the novation point for consultants.

Formerly, the consultants were allowed to develop design and documentation up to 70-80% in the pre-novation period. Now, novation occurs at the minimum point in design development. That is, for each discipline such as architectural, civil, structural, mechanical, electrical, landscaping etc., the scope of work is identified to the extent that the work can be quantified for pricing by tendering contractors.

This approach may be called 'early' novation. It inevitably has an effect on pre-novation consultants' and sub-consultants' fees. Further, the balance between pre- and post-novation fees changes to reflect the less-advanced state of the design at novation.

Consultants program progressive design and documentation in such a way as to allow sufficient time for the owner to review before construction tenders are called. The following approximate review points occur:

- The brief establishment.
- The 5% review (hold point) solely for architectural work. This consists of site master planning and conceptual site layout, conceptual floor plans, design philosophy report and budget cost plan.
- The 15% review (hold point) for architectural drawings, engineering drawings and elemental cost plan.
- The construction tender review for architectural, engineering and detailed elemental cost plan at novation point.

Generally, at pre-novation, the work required to be completed to enable tender documentation to be priced by the contractor consists of drawings, specifications, schedules, and lists.

At post-novation, the consultants complete all design and documentation commonly during the first or the second month of the construction contract period, and they issue construction drawings sufficient for the contractor to proceed with the construction.

The question is: when is the most appropriate time for the novation to occur? There are many arguments and different interpretations to this issue.

One suggestion is that the extent of design development at the point of novation may accord with the summary in Table 4.4 for each discipline.

The reason behind this early novation is to reduce the pre-novation duration by calling for construction tenders as early as possible. Consequently, the benefit to the owner is a reduced total duration for documentation and construction.

EXERCISE

1. For early novation, documents are provided at somewhere between 15-25% completion, which is regarded as a minimum point. What problems might be encountered with only 15% complete at the time of novation?

2. In some cases, once practical completion of the work has been reached, the consultant reverts to being employed by the owner to prepare the list of defective and outstanding items, monitor the completion of the same and certify the completion of the project at the end of the liability period. Payment for this section of the consultant's work is made by the owner. Is it reasonable for the consultant to switch from owner to contractor and back to the owner, without a conflict of interest?

3. When early novation was first introduced to the consultants, it was hard for them to accept. There was a lot of uncertainty for the consultants and their various discipline sub-consultants in identifying when would be a suitable time to finalise documentation for tenderers. Most consultants were willing to prepare the tender documents as early as possible, but they believed using early novation involved a substantial risk which they would like to avoid. Are their fears well founded?

4. Early novation was found to be not supported by the majority of the consultants, or favoured by

Table 4.4. Summary for ech discipline.

Discipline	Novation point
Architectural	Nominally 25% complete, comprising review drawings with minimum dimensions added, i.e. structural grids, overall plan dimension and sections, heights, with windows elevated, doors and door hardware scheduled, internal partition/wall types, finishes and external wall finishes scheduled.
Structural	Nominally 25% complete, comprising footing and floor slab sizes and re-inforcement, structural steel sizes, including cladding grids, wall framing and roof purlins shown on structural framing plans, elevation and sections.
Civil and external hydraulics	Nominally 50% complete, comprising developed site plan with paving areas determined, stormwater and drainage disposal established with pipework sized. Floor levels and paving levels shown, paving types scheduled.
Internal hydraulics	Nominally 15% complete. Pipework layouts are not provided. Drawing and functional design brief requirement coordinated, with sanitary fixtures scheduled.
Mechanical	Nominally 25% complete, with air conditioning performance data scheduled, extent of air conditioning system shown on drawings with ductwork runs established and coordinated. Special air conditioning requirements to be scheduled in compliance with the functional design brief.
Electrical	Nominally 25% complete, with power and lighting requirements scheduled on the drawings. Main switchboard and distribution boards shown, cable sizes determined, and single line switchboard diagrams shown. Special fixtures scheduled for both power and lighting.
Fire	Nominally 15% complete, with performance criteria established and extent scheduled.
Landscaping	Nominally 15% complete, with extent of landscaping areas determined.
Master specification	Specific description of materials and workmanship for all disciplines.

contractors. Contractors preferred the consultants to continue to provide them with the concept design and the specification, and would rather support the develop-and-construct technique.

The majority of contractors and consultants were more familiar with design-and-construct contracts. When it came to early novation, the consultants and sub-consultants had difficulty because of a lack of appreciation of the process, and lack of previous experience. Is acceptance of early novation an education issue, or is the small percentage of work complete at which novation occurs the issue?

5. When novation occurs is the same argument as to what proportion of design should the owner do in the design-and-construct method. From 0% to 99% is possible. When does the owner feel confident that its requirements will be met by the design-and-construct contractor? How far does the owner have to develop the drawings and the specification in order to achieve this? On what else besides the owner, contractor and type of work does the changeover point depend?

4.5.2 CASE STUDY – STORMWATER RETENTION BASIN

Introduction

During the certification process, for deferred work being completed by the contractor, an incident occurred where the rights and obligations placed upon the various parties under the design novation agreement were questioned, and subsequently tested by means of a claim. The particular situation involved an inadequate design, which had been novated to the contractor.

The design, that was lacking, had been undertaken prior to the signing of the contractor's agreement. This was done to facilitate early and timely construction. The owner had let the design work out to various consultants. The designs, novated to the contractor, were considered to be 'near final'.

To enable the contractor to legally fulfil this role, a design novation agreement was created and implemented for all work milestones, where the owner had already commissioned the designers. The agreement was between all three parties.

The incident

The incident which occurred and gave rise to the questioning of the design novation agreement indicating rights and obligations, involved a stormwater detention basin. The basin formed an integral part of the trunk drainage system in the catchment area.

The basin was designed and constructed as part of the original Stage 1 contract. Due to the staging of the trunk drainage works for the catchment, it was determined that additional temporary measures were required to improve the water quality until a more permanent structure was constructed upstream. To achieve this requirement, a temporary weir, constructed across the outlet to the basin was deemed to provide the necessary extended detention required to improve the stormwater quality.

The owner subsequently commissioned the design consultants to design the temporary weir structure and an outlet, to provide the necessary detention time. The designs were completed and issued to the owner and appropriate government department for comment. Comments were received and issued to the designer for its consideration.

The drawing showed a 2 m high wall located at the entrance of the culvert between two wingwalls. At the base of the wall was a 90 mm diameter uPVC pipe, which allowed the stored volume of water to drain in 24 hours under controlled conditions. To prevent any blockage of the 90 mm pipe, an additional vertical slotted pipe wrapped in geotextile fabric was attached. Behind the wall, the existing gabions remained to provide scour protection to the entrance of the culvert.

The work was constructed in accordance with the design plans and specifications. Both the overseeing government department and the owner confirmed that they were satisfied the work had been constructed in accordance with the design and were in the process of approving the contractor's payment claim when the catchment experienced two significant storm events. This represented the first rain in many months, and more importantly the first since construction on the wall had been completed. In hydrological terms, the storms were considered no more than a 1 in 1 year or maybe a 1 in 2 year event. During the initial storm event, the volume of runoff into the basin was such that the wall was overtopped.

An inspection of the wall later, showed no evidence of any structural damage, however it was noticed that a significant number of relatively large rocks in the gabion behind the wall, had been disturbed. During the next few days no significant rainfall occurred, however the rate of outflow from the basin indicated the retention time far exceeded the designed 24 hours. The owner flagged the issue, and the contractor and the designer promptly addressed the issue and suggested that the problem with the drainage times was due to the geotextile fabric becoming blocked due to the higher than expected silt loads. Once the fabric was removed, some improvement in the outflow was noticed, however not until the owner instructed the contractor to totally disconnect the vertical pipe, was the outflow consistent with a period of 24 hours.

Sometime during the following week a second rainfall event occurred. The volume of water was such that the wall was overtopped by as much as 500 mm. On inspection after the event, it was found that the gabion had been totally dislodged and was washed some 50 m downstream. At the culvert entrance, scouring of the order of 750 mm had occurred, and this now was threatening the structural integrity of the culvert. Constant surveillance of the basin over the next week showed the basin emptying in a period approaching 2 weeks after rainfall.

The claim

At this point, the owner concluded that the work was not fit for purpose, and did not meet the functional requirements of the original brief. The owner subsequently advised the Project Certifier to suspend the payment process until rectification works had been completed. This effectively placed the owner and the contractor in dispute.

The contractor and the designer both agreed that the problem with the gabion was of an extremely critical nature. Subsequently the design and construction of a reinforced concrete slab was immediately carried out which ensured the integrity of the culvert, should further rainfall occur. (The issue of the poorer-than-expected outflow was not addressed. This was another component in the dispute.)

The contractor, in the belief that it had completed the original work in accordance with the design, requested the payment process recommence and that it be reimbursed for the cost of rectification work, the basis of its claim being that this additional rectification work was outside the scope of its contract. It claimed that it was not responsible for a design completed by external parties and approved by the owner and the overseeing government department.

The owner dismissed the claim on the basis that the design novation agreement transferred all rights and obligations in regards to the design to the contractor. The owner pointed out that the design in its view should have considered the effect of overtopping on the gabion, and modified the structure accordingly. The owner also put forward the fact that all drawings were passed to the overseeing government department and the owner for comment, not approval. The owner relied on the contractor and the designer for the expertise in the area of concern. The owner did concede to recommencing the original payment process given that repairs had been effected.

The contractor subsequently claimed against the designer for rectification costs due to an inadequate design. The designer in turn dismissed the contractor's claim and invoiced the contractor for the cost of the additional design work. The designer's reason for rejecting the contractor's claim was based on the fact that the original brief called for the design of a temporary weir structure. Therefore, given the structure was only ever to be temporary, the design brief implied a necessity to minimise the cost of the structure. Therefore, replacing the gabions was considered uneconomical for temporary works. It is debatable whether these considerations were ever given.

The dispute continued with the contractor emphasising a different line of reasoning – if the contractor and the designer had recommended the removal of the gabion in the original design, the owner would have been willing to pay the higher price. Therefore, paying for the rectification costs now should be no different, in terms of the argument, than paying then.

EXERCISE

1. The case has brought into focus the way, in a particular design novation procedure, various opinions will still exist. It also focuses attention on some perceived inconsistencies between the original intent of the novation agreement and the agreement signed in this case.

   The inconsistencies relate to the actual content of the novation agreement. In this dispute, the questions being asked were:
   - Does the owner, having engaged and paid the design consultant prior to the novation, have any responsibility for faults or oversights in the design?
   - Was the owner entitled to undertake the novation, given the design was completed prior to the agreement and all design fees had been paid?
   - Did the contractor fully appreciate the terms and conditions of this design novation agreement and its associated responsibilities?
   - Was the contractor placed in a difficult position, given the designer had been paid in full, and this prevented the contractor from maintaining a usual contractual relationship with the designer?
   - Did the designer have any ongoing responsibilities to the contractor, given the design was completed prior to the novation?

   Was there a fundamental flaw in the nature of the agreement signed by all parties? Does this suggest any agreement between the contractor and the designer may be unenforceable?
2. Do you believe the contractor has entitlement/no entitlement to rectification costs due to the novation agreement based on the above information?
3. Is the standard design novation agreement applicable in this case where all the design is complete, rather than part design?
4. One trouble with new contract conditions such as this one-off contract and involving design novation, is that the 'water' has not been tested in the courts. How do you deal with the situation where you don't know the way the courts will interpret the contract?

4.6 DESIGN-CONSTRUCT-AND-MAINTAIN

Where maintenance issues are important, or where the maintenance costs are important in the total life cycle costs of a facility or asset, there may be a need to make the contractor accountable for its performance in the design and construction phases, through having the contractor take responsibility for maintenance (in addition to design and construction). The contractor then optimises its approach over the combined design, construction and maintenance phases, rather than only over the design and construction phases as in D&C.

[An alternative is to use design-and-construct, and have the contractor warrant at tender time that the maintenance costs of the end-product will not exceed the contractor's tendered estimate of maintenance costs, over a defined time period. However, enforcement of this might be difficult.]

Risk associated with the maintenance is transferred from the owner to the contractor, while the owner maintains control of the facility.

The period for which the contractor is responsible for maintenance, after construction is completed is discretionary, and depends on the facility or asset. It may be ten, twenty or more years.

Design-construct-and-maintain (DCM) is distinguished from BOOT type commercial developments, in that the contractor never owns the facility or asset in the former, although in each there is a responsibility for maintenance.

The contractual linkages are the same as in Figures 4.12 and 4.13, but with the contractor's scope of work enlarged to additionally include maintenance.

A number of major highways, water treatment facilities, petroleum infrastructure and like facilities have been delivered using DCM.

Example

When a road authority has a road constructed, it includes a 10-year maintenance period in the contract. The total design-construct-maintain deal creates a sense of ownership in the road by the contractor.

Why DCM?

With conventional contracts, there is commonly a defects liability period, during which the contractor is able to rectify any problems with workmanship or materials. This period is commonly one year. Should defects arise after the expiry of this period, it can be difficult to get the contractor to remedy them. Contractors may only remedy such defects under commercial pressure, or for image or reputation reasons.

As well, there is the maintenance of the facility or asset ongoing during this defects liability period and beyond.

In conventional design-and-construction delivery, there is little motivation for the contractor to provide anything other than the minimum requirements, and certainly there is no motivation to consider an optimum life cycle costing approach to design and construction activities. It may be of little concern to the contractor if, like the 'one hoss shay', the facility is unusable on day one after the expiry of the defects liability period. The contractor essentially washes its hands of the facility at the expiry of the defects liability period. The contractor is able to use the cheapest and lowest acceptable quality materials, labour and design which satisfy the contract.

Traditional delivery, and deliveries where the owner keeps control of the design, are able to address this partly, through the owner overseeing the design and specification aspects. The design and specification, in such cases, should reflect the owner's long term interest in the facility.

DCM issues

The important differences between DCM and D&C and traditional contracts are as follows:

- The defects liability period is as long as the maintenance period.
- The DCM contractor (during the maintenance period) cohabitates with users and the owner of the facility.
- Remuneration for the maintenance period should not be on a cost-plus basis, otherwise there is no motivation for the contractor to think about maintenance issues during the design and construction phases. Commercial motivation of the contractor will depend on the payment arrangement during the maintenance period. Lump sum payments are recommended, commensurate with the extent of risk that the contractor is being asked to shoulder, acknowledging the difficulty of being able to predict precisely future maintenance requirements.
- While a subcontractor may undertake the maintenance work, the contractor would be expected to be responsible overall for the performance of the subcontractor, and not be expected to opt out of this responsibility.
- The owner's brief to the DCM tenderers needs to make reference to life cycle costing matters, and in particular solutions that minimise life cycle costs for the owner, while still having a facility of an acceptable standard. This may mean, for example, tenderers developing more expensive designs in return for lower maintenance costs.
- The owner's brief should not be overly prescriptive, else innovation may be stifled. The scope of maintenance needs definition in order to be priced competitively by tenderers, and to avoid subsequent adjustment of the maintenance payments for matters not thought of by the owner at the time of tendering.
- Assurance is necessary that the contractor will remain in business for the duration of the maintenance period.

DCM providers

DCM contractors may elect to establish maintenance divisions within their organisations, and search for other maintenance work as well, or they may elect to subcontract-out the maintenance. That is, the maintenance may be carried out in-house or outsourced by the contractor.

Alternatively, a specialist maintenance company may outsource the design and construction work. The designer and constructor become subcontractors to the maintenance company, or a construction manager is engaged to deliver the construction through trade contractors.

4.7 PROJECT MANAGEMENT

When reference is made to the project management method, reference is commonly to Figure 4.17. (Some alternative configurations are given later.) Here the project management consultant (a contracting organisation or consultant offering specialist project management services) acts as an adviser to the owner and assists the owner carry out the project to the owner's goals. (Essentially Figure 4.17 is like the *traditional delivery* method /

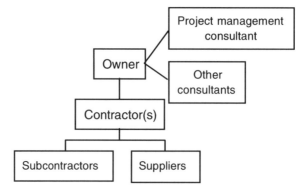

Figure 4.17. Project management method (showing contractual links).

construction management method, with the addition of a project manager adviser to the owner.)

In this project delivery method the owner engages an agent (a project management consultant), to provide a management service for all phases of a project from inception to completion and who, in particular, undertakes, for a fee, the management of the other consultants and the contractor for the project, within the parameters defined by the owner. The project manager is a single point of advice to the owner on all project matters.

Payments to the consultants and contractors can either come directly from the owner, or via the project manager through an account set up by the owner.

In disputes between the contractor or consultants and the owner, and in which there is also a dispute with the project manager, the owner may require the project manager consent to the joinder in the dispute hearing.

The intent of having such a project manager is to bring a business-like approach to the coordination of all project participants and all project work, particularly for large projects. *It highlights the importance of the project management function.* There now exists a consulting industry that provides specialist project management services, whereas formerly professionals dabbled in project management as a secondary occupation to design or construction.

Conflict of interest

There is a *conflict of interest* if the project management consultant is also, for example, a design consultant, and while such dual roles (designing and policing design) are common, it is not recommended from the owner's viewpoint. As well, the owner needs to be aware of the specialist design consultant acting as a project manager, and perhaps contract administrator, without the necessary skills in these later areas.

Project managers, in former days, commonly took on multiple roles. The recommendation today is for a single role.

Payment

Commonly, the project manager is paid a fixed fee. The fee may be adjustable on a bonus/penalty arrangement for cost and time project performance. The fee would be expected to cover the manager's profit and head office overheads. Genuine and approved expenditure incurred by the project manager (including payment to people), in managing the project, would be reimbursed. Expenditure arising out of disputes with the owner, carelessness, or incompetence would not be reimbursed.

Termination of the project manager's services by the owner is possible at any time, with the project manager receiving the appropriate portion of its fee, together with approved reimbursements.

Terminology

There is no universally accepted understanding as to what is meant by the term 'project management'. This lack of an accepted definition is further confused by the fact that the player organisations within the project may each have a key member of their organisations with the position title of 'project manager', and they manage their own organisation's 'project' (which is a subproject of the owner's overall project). Many people confuse the function of project management with the delivery method called project management.

The following list of the project manager's responsibilities and the following comments on the relationships between the project management consultant and the other consultants may assist clarify the meaning of the 'project management method'.

Where several projects are managed within an overall program, this might be referred to as *program management*. And if subprojects are regarded as projects, then the terms 'project management' and 'program management' become interchangeable.

The term 'professional project management' might be used to describe what is called 'project management' in this book.

The notation PM might be used by some people instead of 'project manager', and PPM instead of 'professional project manager'.

Project manager's responsibilities

The project management consultant's responsibilities include:
- Documenting, letting and administering consultancy agreements with design consultants and specialist consultants.
- Developing a master design and specification for the project.
- Preparing and controlling a master program for the project, setting out the times within which the various parts of the project, including all relevant on-site and off-site construction activities, are to be executed. The master program should include the dates by which information and decisions are required from the owner, design consultants, specialist consultants, authorities and others involved in the project.
- Preparing and controlling a master cost plan setting out all relevant cost analyses, budgets, cost control systems and the like.
- Preparing and controlling a master rate of expenditure plan setting out rates of expenditure and the requirements of monetary funds for the project.
- Developing and implementing an industrial relations and safety program dealing with matters such as the location and type of amenities for the project workforce, the com-

munication framework with union and safety representatives, the basis for the prequalification of potential direct contractors, dispute resolution procedures and the like.

- Developing and implementing a quality assurance program with respect to both the design and the construction for the project.
- Preparing contract documentation for the contract with the contractors, incorporating the designs and specifications prepared by the design consultants.
- Entering into a contract with each contractor, the project management consultant acting as disclosed agent of the owner.
- Administering the contract with each contractor on behalf of the owner.
- Arranging for certain common user facilities such as cranage, scaffolding and the like.
- Reporting regularly to the owner on all aspects of the project relevant to the management agreement.   (based on NPWC/NBCC, 1990)

Project manager's relationship to others

The project management consultant's relationships with the owner and other consultants may be summarised as follows:

- Under the project management method, the project management consultant enters into a project management contract with the owner and is responsible to the owner for managing both the design and the construction for the project.
- However, unlike the contractor under the design-and-construction method, the project management consultant is usually engaged on a fee-for-service basis and does not assume the time and cost risks associated with the design-and-construction method. The risk allocation depends upon the particular requirements of the owner and the project.
- The project management consultant, acting as disclosed agent of the owner, engages the design consultants and specialist consultants required for the project.
- The owner has a contractual relationship with the consultants. The project management consultant would normally be responsible to the owner for the co-ordination and control of the other consultants' work but would not normally be responsible for the adequacy of the other consultants' work.   (based on NPWC/NBCC, 1990)

Advantages to the owner

- Owner responsibilities are reduced through having trust in the project manager's skills.
- *Project management ensures that the project development process is more cost and time efficient.*
- *The [owner] may replace the project manager without causing the project to slow down or stop (as would be the case when the general contractor is dismissed or goes bankrupt).*
- *The administration of the contract is done in a more professional manner. This should reduce the level of contractual disputation.*

Disadvantages to the owner

- The owner may wish to retain financial control and/or avoid paying the handling margins to the manager.

- *The project manager may not always be an ideal party to deal with industrial relations issues.*
- *The [owner] must appoint the project manager early in the project. This may not always be convenient to the [owner].*
- *The success of the delivery method will depend on the ability of the individual project manager, hence selection is very important.*
- *The project manager will require to be given functional control of resources to be fully effective.*
- *The [owner] carries the greatest share of financial responsibility of the project and other major risks.* (Uher & Davenport, 1998)

4.7.1 ALTERNATIVE CONFIGURATIONS

A number of alternative configurations for project management, or modifications to the above configuration, may be seen.

Project management in conjunction with the traditional method

The contractor is engaged in the traditional delivery mode (Fig. 4.18).

Project management in conjunction with design-and-construct

The contractor is engaged in one of the various design-and-construct modes (Fig. 4.19).

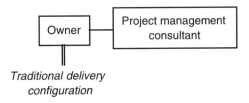

Figure 4.18. Project management method (showing contractual links) – alternative.

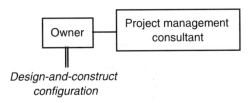

Figure 4.19. Project management method (showing contractual links) – alternative.

Project management in conjunction with construction management

Trade contractors are engaged through the assistance of a construction manager (Fig. 4.20). The project manager has overall responsibility for the project, while the construction manager has responsibility for construction only, and reports to the project manager.

In some cases, the project manager and construction manager are one, and the distinction between project management and construction management delivery methods disappears. Generally, though, a project manager has a longer time involvement in a project than a construction manager, who typically is only concerned with the construction phase of a project.

Such a delivery method might be called an *integrated contract*, or the term 'commercial project management' might be used. The notation CPM might be used by some people instead of 'commercial project manager'.

Consultants contractually linked to the project management consultant

If the design consultants and specialist consultants (Fig. 4.21) are engaged as subconsul-

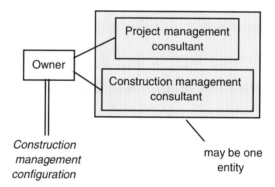

Figure 4.20. Project management method (showing contractual links) – alternative.

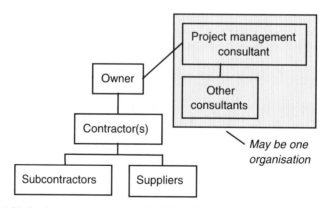

Figure 4.21. Project management method (showing contractual links) – alternative.

tants to the project management consultant, the owner has no contractual relationship with the consultants and the project management consultant is responsible to the owner for the adequacy of the other consultants' work.

The precise nature of the relationships between the project management consultant, the owner and the other consultants would depend upon the terms and conditions of the various consultancy agreements between the parties.

Certain project management consultants may have in-house design and specialist resources and, with the owner's approval, the project management consultant may elect to use its own resources rather than engage external consultants. That is, on some projects, one consultant acts as project management consultant coordinating the work carried out by the other consultants, while also taking on a design role. This can result in a *conflict of interest*.

In-house project management skills

Project managers may, of course, be located in any of the boxes drawn in Figures 4.17. For example when an owner has in-house project management skills, the project management box comes inside the owner's box.

Project manager assumes the risk in the project outcomes

An alternative arrangement, where the project manager assumes the risk in the outcomes of the project is shown in Figure 4.22, but this is closer to the managing contractor arrangement, and is generally not favoured by project managers.

It has appeal from an owner's viewpoint as it provides a single point of responsibility. Should something go wrong with the project, the owner need look to one entity for a remedy instead of the many entities involved in the earlier project management arrangements.

The project manager could be engaged on a cost-plus basis, e.g. guaranteed maximum cost with a cost-saving arrangement. The appointment may be with brief documentation, which would be developed by the project manager.

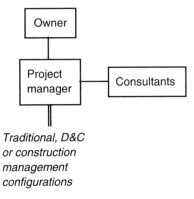

Figure 4.22. Atypical reference to project management method (showing contractual links).

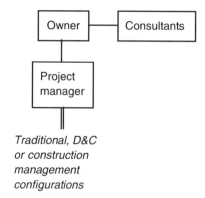

Figure 4.23. Atypical reference to project management method (showing contractual links) – alternative.

A change on this is where the consultants contract directly with the owner

The owner retains the relationship with the consultants (Fig. 4.23). The owner then manages the interfaces between the design and documentation and construction. The owner, or its consultants, prepares the concept design and detail design and documentation for the work.

This has disadvantages for the project manager if the project manager is responsible for final costs and accept risks inherent with these conditions, for example where the consultants are free to spend money as they see fit, rather than in the best interest of the contractor or within estimated project costs. Difficulties and conflicts arise from the project manager not having control of the consultants and liaising with them via the owner. (However, if the owner accepts responsibility for design and documentation, and adjusts the final cost as changes in design occur, then the project manager is not disadvantaged.)

4.7.2 CASE STUDY – CAN-MAKING

Installation of a third can-making line to supplement two existing lines

The delivery method employed by Canns Company for this project was somewhat unique. Factors which made this project unique were:
- All equipment was sourced from overseas.
- One of the successful tenderers was asked to coordinate the overseas component of the project.

An aluminium aerosol can-making line consists of many processes:
- Impact extrusion.
- Trimming and brushing.
- Washing.
- Internal lacquer spray.
- Curing oven.

- Decoration.
- Decoration curing.
- Neck and dome forming.
- Strapping and bundling.

For each process, there may be up to four different manufacturers offering different technologies, performance and price, and an aerosol line may be built using any combination of different manufacturers' equipment.

A project manager was appointed to work for Canns specifically to determine which company/manufacturer to use for each process and to handle all contracts. The project manager was appointed with specific knowledge in this field.

All manufacturers of the equipment were asked to submit a tender to firstly supply the specified equipment. The selection process then began, and was based primarily on the technological advantages of one piece of equipment over another. Each tender was put forward on a lump sum contract basis.

Initially it was thought that one manufacturer could supply the entire can-making line, and thus would be responsible for all costing, commissioning and guarantees. Although this was possible, it was considered desirable to choose technology from more than one manufacturer. A total of three manufacturers were finally chosen. As all processes were linked together, it was essential that complete co-ordination and meshing of these companies existed to make the project work.

Company X, whose equipment was chosen for six of the nine processes, was asked to take authority and control for the mechanical aspects of the entire combined process, which included the other two equipment manufacturers (Companies Y and Z). Companies X, Y and Z were all based overseas.

Company X was responsible for ensuring that all processes fitted together and performed to guarantees, completion to a set time and a budget. Company X was then to be paid an additional fee, on top of the initial equipment contract amount, for managing the project. This method of procurement was chosen because all equipment was made in Europe, and therefore coordination from an overseas point of view was preferred. The other successful manufacturers were then asked to co-operate with Company X as part of their contracts, and they both agreed (see Fig. 4.24). All equipment specifications were known, and therefore lump sum contracts were used.

Additional to the equipment supply from overseas, factory modifications and an upgrade of services (air, water, electricity) were necessary in order for the third can-making line to be successfully installed and commissioned.

Factory modifications consisted of office relocations, storage relocation and the introduction of climate control. Tenders were requested from local companies for this work. After a company was chosen to perform the work, it was responsible for all modifications, based on a lump sum contract. The successful company was also to employ a project manager. The upgrade of services was contracted to local specialists in each individual area.

The role of the project manager at Canns was to ensure that overseas equipment was working harmoniously and that the factory modifications and the services upgrade were all tied together. Cost accounting for the entire project was carried out using the company employees.

All work was contracted instead of using in-house resources. The primary reason for

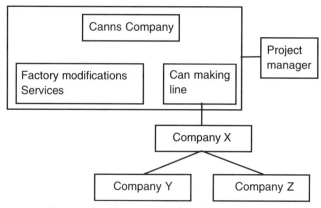

Figure 4.24. Contractual links, Canns Company.

this was that, as a company, Canns was in the business of making cans, not manufacturing and developing new technology or building modifications. The company specialists were in can-making. Maintenance of all new equipment was to be done in-house, as was current practice at the time. Additionally, in-house labour was to be utilised in the production aspects of can-making.

EXERCISE
1. It was found that there were many companies who specialised in all areas of work needed. Does this point to a particular delivery method?
2. Much time was spent in preparing the scope of work for all of the contracts. It was considered harder to make changes once the contract was in place. Additionally, much resource re-allocation, mostly derived from within the company, took place to prepare and administer the contracts. What happens to these staff after the project is completed?
3. The project manager was hired to perform overall co-ordination and for the preparation of contracts. The project manager was on a contract for a set period which included a two year period after the project was completed. Why do you think such an arrangement was chosen for the project manager, and why was the project manager outsourced?
4. The concept of hiring a project manager to work within the company was an attempt to keep control of the entire project (equipment, factory modification, services upgrade). The project manager was accountable for the entire project. Could this goal be expected to be achieved?
5. There were multiple contracts involved in this project – overseas and local components. Would using a single contract have been possible?
6. Initially, all contracts were of the lump sum type. However, Company X was then to be paid an additional fee for overheads and management. With the lump sum contracts for Companies Y and Z, the currency exchange rate was included as the only variable. Although Company X was responsible for the mechanical co-ordination and meshing, Companies Y and Z were still ultimately responsible to Canns. What concerns to Canns are involved in this approach?
7. Using Company X involved dealing with a new, overseas company with a different culture to Canns. What are the risks in using a variable fee arrangement here? Could it give rise to extremely large fees? Also, is it possible that Company X may take advantage of this?
8. How could the risks associated with the overseas factor, and the less than desirable communication links be handled? What place do formalised procedures that detail regular reporting back, or updates to the owner have?

4.8 DESIGN AND MANAGEMENT

EPCM is an acronym for *engineering, procurement, construction and management*, a commonly used form of consultancy contract provided by engineering consultants. It provides for the provision of engineering design, project management, procurement, contract administration and management, and construction supervision and management.

EPCM is an extension of, what is described in this book as:

- 'Project management in conjunction with construction management' .

or

- 'Consultants contractually linked to the project management consultant'.

But some writers give alternative arrangements.

The consultant takes on the design role additional to the management role. The owner assumes the risks associated with the construction outcome.

4.9 CONSTRUCTION MANAGEMENT

In this project delivery method (Fig. 4.25), the owner engages an agent (a construction management consultant – usually an industry consultant or contractor organisation or individual with a knowledge of and experience in construction), to provide a service for the construction phase, normally provided by a general contractor, particularly related to the control, management and co-ordination of the construction for the project. The owner engages the design consultants and specialist consultants for design and documentation. The construction manager offers pre-construction advice to the owner, and may also liaise with the design consultants to ensure constructability and construction issues are addressed at the design stage.

Such a delivery method might be termed an *agency* construction management *arrangement*.

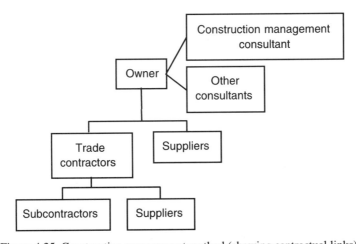

Figure 4.25. Construction management method (showing contractual links).

Terminology

The terms 'construction management' and 'project management' are, to many people, the same thing. Occasionally the expression 'construction project management' is used and this further blurs the distinction. The terms are abused by many. This chapter clearly distinguishes between what is the project management delivery approach and what is the construction management delivery approach. Note also, that the terms are used in industry in a number of other senses not connected with contractual issues.

The term 'professional construction management' might be used to describe what is called 'construction management' in this book.

Some people may refer to this form of delivery as 'project management', and this highlights loose lay usage of the terms 'construction management' and 'project management'. In some cases, the project manager and construction manager are one, and the distinction between project management and construction management delivery methods disappears. Generally, though, a project manager has a longer time involvement in a project than a construction manager, who typically is only concerned with the construction phase of a project.

The notation CM might be used by some people instead of 'construction manager', and PCM instead of 'professional construction manager'.

Trade, specialist, independent or 'separate' contractors

Going hand-in-hand with the delivery method, usually the work is broken down into packages suitable for different trade, specialist, independent or 'separate' contractors. Each package can be configured differently in terms of payment and other conditions. Each trade package must stand on its own. The construction manager does not undertake any construction work, but rather manages others in the best interests of the owner. Project risk is with the owner, not the construction management consultant. Trade contractors contract directly with the owner.

(It is possible for the construction manager to undertake some construction work, if the necessary skills exist. However, this introduces an arrangement the same as the traditional delivery method. The construction manager may provide some project or site facilities used in common by the trade contractors, as well as clean-up services, temporary works, works protection, and 'papering over the cracks' where different trades interface.)

With multiple trade contracts, the owner can call for tenders progressively, and can pace and control the work as desired.

Performance of the manager

The [owner] can sue the construction manager for not managing as agreed but the measure of damages is not the cost of fixing the defective work carried out by the separate contractors. It is the amount necessary to put the [owner] in the position in which the [owner] would have been had the construction manager properly managed.

(Uher & Davenport, 1998)

Should the construction manager not perform, the owner can terminate the manager's services, while still keeping in place the trade contracts, and progressing the project. A termination clause may be included in the contract between the owner and construction manager. Trade contractors can be notified of the owner's new agent. Dismissal is the usual sanction against non-performing construction managers. Liquidated damages (payable to the owner) may apply for the trade contractors, but not the construction manager.

Payment

Payment to the construction manager is commonly a fixed or percentage fee. The construction manager's fee is intended to allow for its profit, head office overheads, management personnel, and the cost of running disputes with the owner or subcontractors. Guaranteed maximum cost, and target cost are also used.

The owner reimburses the construction manager for the project direct and indirect costs incurred by the construction manager (e.g. infrastructure services such as power and water), though it is possible to incorporate all of this in an enlarged fee.

The fixed fee tends only to be suitable where the scope of work can be well defined before the management contract is formed. In such a case, the traditional delivery method would be a viable alternative. Where the scope of work cannot be well defined up-front, the fixed fee would have to be adjusted later on anyway, unless a prior upper limit on project cost and time had been established and the project was still within these limits.

Because of the smaller size of the trade packages, the specialist contractors will usually be engaged on a fixed price basis, without provision for rise and fall.

The owner pays the specialist contractors, either directly or through the construction manager for work certified by the construction manager. In the latter case, the money is held in trust by the construction manager until payment occurs, and cannot be used by the construction manager for other purposes. In both cases, the construction manager certifies the amount payable.

The owner retains control over the security and retention moneys from the trade contractors. Were these to be held by the construction manager and the construction manager became insolvent, the owner would still have to account for the money.

Insolvency of the owner leaves the construction manager without its fee, and the trade contractors without payment. There is no liability on the construction manager to pay the trade contractors. If the owner is a public sector body, where insolvency wouldn't be expected, the trade contractors are protected from insolvency of the construction manager (compared to a non-agency arrangement – traditional or managing contractor arrangement – where they may not be protected).

Services provided by the construction manager

The construction manager administers the contracts with the specialist contractors on behalf of the owner, as an agent of the owner.

Services provided by the construction manager include:
- Recommending practices to the owner, including pre-construction and design advice; liaising with designers.
- Estimating.
- Tenders, tender evaluation, contract documents, contract administration (including variation approval).

- Finance – budgets, monitoring, control, reporting (commonly on a committed cost basis for each package), forecasting, cash flow.
- Scheduling/programming, monitoring, control, reporting, forecasting.
- Procuring materials, equipment; purchasing, expediting; hiring labour.
- Approvals from government authorities.
- Planning site facilities; provision of site facilities.
- Managing industrial relations, safety.
- Insurances.
- Coordinating trade contractors.
- Project meetings, handling disputes.
- Inspections, quality.
- Records, administration; supervision, payments.
- Security.

A reduced-services version of construction management is for the owner to employ a contract or *contracts management consultant* (one who solely manages contracts).

Conditions of contract

The construction manager would typically be engaged based on a performance specification and expected role during the design and construction phases, rather than on a detailed day-to-day work outline.

The conditions of contract for the trade contractors would be little different to that used for the contractor under the traditional delivery method, except with the addition of coordination and cooperation requirements.

Design

Sometimes, construction managers are people or organisations with a predominantly design background, and there can be the temptation to have the one group look after both construction management and design.

It is possible for the construction manager to undertake some design, if the necessary skills exist. In some cases it may be unavoidable that the construction manager is called upon to fill some gap in the design, or to design some temporary works. However, this is not recommended because it places the owner and construction manager into an adversarial relationship, a relationship the construction management delivery method is trying to avoid. The delivery method is intended to remove the pressure to make a profit through securing some trade advantage from the owner or specialist contractors.

With the construction manager also providing design, there is the potential for conflict of interest. There is the potential for a specialist contractor to blame the construction manager's design for any problems encountered, while the construction manager can put the blame on the specialist contractor as a cover up for any design deficiencies.

Where the construction manager provides design, it can be difficult for the owner to terminate the construction manager's services quickly, should the construction manager not be performing as desired.

Comparison with traditional delivery or managing contractor (without design) delivery

The conditions of contract in construction management or managing contractor refer only to management services.

Because there is no contractual link between the construction manager and the trade contractors, but rather only between the owner and the trade contractors:

- The construction manager is immune from contractual claims from the trade contractors; the cost to the construction manager of running disputes with specialist contractors is not present; the risk associated with disputes is not present; reduced construction manager's fee.
- The construction manager cannot sue the trade contractors for non-performance.
- The trade contractors are protected from insolvency of the construction manager.
- With insolvency of the owner, the construction manager has no payment responsibilities to the trade contractors.
- The owner must sue the trade contractors directly, rather than the construction manager, for non-performance; the trade contractors must sue the owner for payment.

The inability of the [owner] to sue the managing contractor when a separate contractor defaults, is seen by some as a reason for not using the agency agreement. Others see it as an advantage because it avoids the adversarial role that exists in the traditional contract. In the agency agreement, the managing contractor can fearlessly uncover defects by a separate contractor. In a 'traditional arrangement', the main contractor is loath to draw to the attention of the [owner] any defects for fear of liability.

The most important feature of the agency agreement is that the [owner] has on the [owner's] side an experienced contractor to assist the [owner] and look after the [owner's] interests.

(Uher & Davenport, 1998)

Advantages to the owner

- Suitable for: projects with only partial documentation or where the owner's requirements become refined with time; situations where disruption or additions/deletions are expected; projects which are complex in nature (e.g. those involving maintenance of user operations); use on default of the contractor under the traditional method.
- The owner does not need to maintain full-time construction staff, but rather engages a construction manager on a needs basis; expenditure is under the owner's control.
- Owner responsibilities are reduced through having trust in the construction manager's skills.
- Small trade packages permit a bigger field of potential tenderers; specialist contractors asked to carry only a portion of the project risk.
- Owner can let trade packages to suit own budget, cashflow and timeframe; phased construction and fast-tracking possible; control over expenditure possible; continual review and refinement possible; cost savings possible in a falling market.
- With no head contractor to act unethically or incompetently, the trade contractor prices should be free of associated allowances.

- *It gives the [owner] flexibility in terms of hiring and firing of construction managers and [specialist] contactors.*
- *Because the construction manager becomes a member of the design team from the start, the construction manager's expertise on construction costs, time and techniques can be of great benefit to the [owner] in the design stage.*
- *Special construction skills may be utilised at all stages of the project, with no conflicts of interest between the [owner] and the manager.*
- *Greater flexibility is possible, enabling the [owner] to take advantage of changes in technology, economic conditions and sophisticated equipment.*
- *By packaging the project into separate contracts, the need for provisional sums or provisional costs is virtually eliminated.*

(Uher & Davenport, 1998)

- *Independent evaluation of costs, schedules and overall construction performance, including similar evaluation for changes or modifications, helps assure decisions in the best interest of the owner*
- *Full-time coordination between design and the construction contractors is available*
- *The ... construction manager approach allows price competition from [specialist] contractors akin to the traditional [approach]*

(Barrie & Paulson, 1992)

Disadvantages to the owner

- The construction manager does not guarantee the project's cost, duration or quality.
- Owner has to have the skills to manage the interface between the designer and the construction manager.
- How competitive/non competitive the approach is, compared with the traditional method, is not known for any given project.

- *... an estimate of the total project cost is not known until the last trade package is let. Early completion may not provide a sufficient trade-off for this risk.*
- *Commitment by the [owner] of expenditure on professional fees prior to knowing his full commitment.*

(Uher & Davenport, 1998)

- *Considerable reliance placed on the [construction] manager to control, manage and co-ordinate the design and construction interface; limited capacity to pass on time and cost risks to the [construction] manager.*
- *A high level of sophistication required of trade contractors to co-operate with the [construction] manager and other trade contractors to co-ordinate construction to minimise interference between the trades.*
- *Hands on continuing involvement of the [owner] in day-to-day decision making for staged packaging, review and refinement of design and cost management.*
- *Commitment of expert [owner's] personnel (or outside consultants) and resources to manage the multiplicity of claims which may flow from direct contracting with a number of trade contractors.*

- *High potential for disharmony and lack of commercial co-operation as between trade contractors.*
- *Blurring of responsibility between trade contractors for defects in construction.*
- *Difficulty in allocating risk for care of the works and thus creating insurance complications so that it may be advisable for the [owner] to itself arrange and control the insurance cover which would necessitate amendments to the trade contracts.*
- *Problems in enforcing timely completion obligations as a consequence of the effects of failure to co-ordinate trade contractors and the difficulty of determining causes of delay as between trade contractors.*

(Department of Defence, 1992)

Owner-builder

The construction management method is the one adopted by people who decide to build or modify their own houses. Such people are referred to as 'owner-builders'. They contract directly with the various tradespeople, while receiving advice from someone (equivalently a construction management consultant) who has knowledge of the building industry. Owner-builders don't have industry building licences, and hence are not allowed to build anything other than their own dwelling, and require the advice of an industry-knowledgeable person (here, a construction management consultant) to make up for their lack of industry knowledge. There are, however, many sorry stories where owners haven't sought or heeded the advice of an industry-knowledgeable person.

Alternative. The term owner-builder might also be used by some people for large organisations who perform part or most of the project work themselves, and engage contractors and subcontractors along traditional delivery lines when needed (Fig. 4.26). Example organisations include developers, traditional public-sector bodies and larger private-sector companies. This usage of the term is different to that adopted above, where a construction management type delivery method is implied.

Organisations using this alternative 'owner-builder' approach would require a relatively large volume of work which is relatively constant over a long time period. There would also be a need to have a clear distinction between project activities and operational

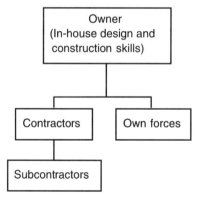

Figure 4.26. Alternative usage of the term 'owner-builder'.

activities within the organisation. The circumstances for adoption are similar to those for using in-house resources (versus outsourcing); advantages and disadvantages of using in-house resources (versus outsourcing) also carry over.

4.9.1 ALTERNATIVE CONFIGURATIONS

The pure management contract is one in which the contractor is made the agent of the owner. All actions of the contractor such as purchasing, negotiation, [trade] contracts, and the employment of personnel are made in the name of the owner. As an agent, the contractor assumes no pecuniary liability. Disbursements and record keeping may be made the responsibility of either the contractor or the owner. Under the agency stipulation, the owner assumes full responsibility for all actions of the contractor. However, by this arrangement, the owner engages the services of a concern possessed of a high degree of technical skill in construction and administration, whose sole interest is that of the owner.

Another type of management contract is that in which the usual independent contractor status is maintained, although the contractor is engaged in a managerial capacity only. Here again, the contractor does not usually carry out any of the construction with his own forces except for its general supervision. He attends to the same duties outlined previously, although his actions are now taken in his own name and on his own responsibility. The contractor pays all bills, generally from funds advanced by the owner for this purpose. The contractor remains responsible for the work until its completion and acceptance by the owner.

(Clough, 1960)

Traditional delivery

Where the construction manager assumes the risk for the construction outcomes, Figure 4.27 applies. If the first construction management delivery method is referred to as an *agency arrangement*, then this alternative is referred to as a *non-agency arrangement*.

Effectively, this is like the *traditional delivery* method, and corresponding conditions

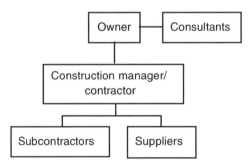

Figure 4.27. Traditional delivery/managing contractor (non-agency) construction management configuration.

of contract would be used, even though the contractor may be called another name, namely 'construction manager'.

The contractual link between the specialist contractors (now subcontractors) and the owner is now through the construction manager, and this has correspondingly changed lines of liability and communication.

With such an arrangement, the owner no longer has anyone managing the construction in the owner's interests, unless the owner or one of the consultants takes on this role. To expect the contractor to also act in the owner's interests as far as managing the construction is concerned, may be asking too much. There is a conflict of interest involved.

Managing contractor

Where the contractor undertakes no construction, Figure 4.27 is essentially the same arrangement as the *managing contractor* delivery method (with the contractor taking no responsibility for design). The construction manager manages the construction on behalf of the owner, yet is contractually linked to the specialist contractors.

The payment basis might be cost reimbursement (fixed or percentage fee), with possibly guaranteed maximum cost, liquidated damages for late completion, and bonuses/incentives for early completion or cost savings. The fee might include head-office costs and profit, or profit only, or there might be a guaranteed maximum cost covering head-office costs and/or project costs.

On a cost reimbursement basis, there is minimal adversary between owner and construction manager. The construction manager acts in the owner's interest.

The construction manager obtains competitive quotes from the specialist contractors. The owner reimburses the construction manager for the amount payable to the subcontractors, as well as the project direct and indirect costs incurred by the construction manager. The construction manager's fee includes its profit, head office overheads, project management personnel, and the cost of running disputes with the owner or subcontractors. The presence of this last item, that is the risk associated with disputes, may mean the construction manager's fee is greater in the non-agency arrangement than in the agency arrangement.

Consider the situation where the subcontractor claims $X, but the owner regards the entitlement to be less than $X. Two options are available to the construction manager – support the owner's view, or support the subcontractor's claim.

If support is given to the owner's view: The subcontractor claims against the construction manager. Should the subcontractor be successful in court or arbitration, the owner will pay $X to the subcontractor, but this still leaves the construction manager with the cost of defending the claim.

If support is given to the subcontractor's view: There is now a dispute between the construction manager and the owner. Again there is a cost to the construction manager in settling the dispute.

If there were a special condition in the construction management contract providing that the [owner] would indemnify the construction manager for [dispute] costs, the construction manager would be protected. There do not appear to be any construction management contracts which do that.

*Other ... costs [to the construction manager] which do not exist under the agency
agreement but do exist under the non-agency agreement [and which would increase the
size of the construction manager's fee] include:*
- *Liability to the [specialist] contractor upon insolvency of the [owner] or delay or refusal of the [owner] to pay an amount due.*
- *Liability to the [owner] for liquidated damages upon delay of a [specialist] contractor.*
- *Liability to the [owner] for defective materials or workmanship provided by a [special-
  ist] contractor.*
- *Liability to a [specialist] contractor for delay costs.*

(Uher & Davenport, 1998)

Such costs, similarly to dispute costs, appear not to be indemnified by the owner in construction management contracts.

Administration

If liquidated damages are present, the associated subcontracts would also be expected to
contain liquidated damages (payable to the construction manager).

The construction manager is accountable for held security and retention money.

4.9.2 CASE STUDY – INSTRUMENTATION PROJECT

A water authority recognised a need to improve how it operated its dams/reservoirs, pipes
and pump stations, to provide potable water for its customers. These facilities had individual control rooms at the various sites linked by telephones/radios. To achieve this improvement, a project was created to address this and other issues related to data gathering,
monitoring and control of the freshwater system.

The water authority engaged a consultant, Mehmet, to progress this into a viable project. The project was defined, split up into tasks/activities, subdivided into regions/areas
and specifications produced. As the main design consultant, Mehmet was responsible for
system design; it advised using remote telemetry units at all installations, linking these together via a communication system and programming a central operations site to oper-
ate/control the system. This central site would enable the entire freshwater system to be
monitored and controlled by a single site all automatically, thus doing away with a number of remotely manned control rooms. Mehmet produced a scope of work and detailed
specification for each facility.

A particular telemetry subcontractor, Rationa, and its licensed telemetry unit was chosen.

A construction manager, Homee, was employed to supervise the site work of installation and commissioning at each facility/site. Tenders were called to carry out this site
work. These were fixed price, detail design-and-construction contracts.

(In addition, the contract to install the communications equipment – to link the sites together – was also let.)

The following description relates to one such detail design-and-construct contract incorporating multiple sites (reservoirs, pump stations, ...) awarded to Motherwell. This involved a 12 month contract for:

- Design – preliminary and final design, as-built details.
- Construction – install instrumentation, civil work, cabling, electrical controls, pipework, testing, commissioning.

The existing system had to be kept operational during this process. The commissioning was only to prove that the instrumentation and wiring connected was correct, and not to test the control back to the central site.

The scopes of work by Mehmet stated what was required for each site. The detailed design would show where to connect the wires, how to mount the instruments, what cable to connect etc.

No variations were allowed, except where it could be shown that the scope of work was incorrect in some way or other.

The construction was managed by Homee, and as such all contact by the subcontractors and suppliers with the water authority or Mehmet was through them.

All design by Motherwell was submitted for approval before construction, but as is common with consultants, Mehmet simply reviewed and did not approve. Thus, any problems with the design were Motherwell's problems.

No as-built drawings, showing the existing details on the sites, were available. This was specifically outlined in the contract, with the scope of work including 'to investigate and verify'. This meant inspecting every site, every circuit to be modified, every plumbing system to altered/replaced, and the incorporation of the actual site details into the design. In most cases, these sites were built a long time ago and had undocumented changes done to them over the intervening years. This represented a particular challenge. Because no 'as-built' drawings were issued to Motherwell, no 'as-built' drawings were required to be issued back to the authority with alterations on them. Motherwell simply issued a complete loop diagram showing the new work performed.

Contractual arrangements

See Figure 4.28.

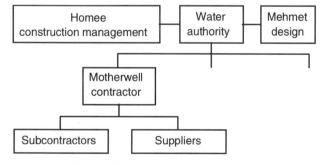

Figure 4.28. Contractual arrangements.

Water authority – Mehmet consultants

A contract to develop design and documentation to address the authority's requirements. A conventional owner-consultant arrangement. Responsible for telemetry unit choice (type, make, model); communication method and requirements; software configuration, standard designs for each site; scope of work for each site (new, modified instrumentation, inputs/outputs required). Commission, prove and validate central control system.

Mehmet was responsible for the overall success of the system. Mehmet reported directly to the water authority.

Water authority – Homee

Homee was employed as the construction supervisor to manage/administer the host of contracts for construction on behalf of the owner. A construction management type arrangement.

Homee could only approve variations of a small dollar value. The owner had decided that the skills of a professional manager outweighed the benefits of close control over the contractors. The water authority oversaw nearly every change, attended all design review meetings and was as well informed as Homee, but did not have to generate the volumes of contractual paperwork to administer the contracts.

Water authority – Motherwell

The water authority contracted with Motherwell as one of a number of construction contractors to install its telemetry equipment and complete the detailed design.

Motherwell was given a partial design and specification, covering what was required, and had to produce the detailed design for owner approval.

Motherwell had no formal link with Mehmet. This allowed competitive tendering that reduced the price of the construction. The disadvantage was that Mehmet, not having done the detailed design, was reliant on the ability of the contractors to alert it to problems. These problems meant, for example, that the contractor could implement the scope of work, but the pump wouldn't start. These peculiar situations relied on the contractor to think about the end goal and not follow the scope of work too blindly. Mehmet had to commission the total system at the end, and so local knowledge had to be relearned. This was inefficient.

As personnel were laid off by one contractor they were employed by the next phase of the project. This included personnel from Motherwell and Homee who were employed by Mehmet, as the work moved into the next phase.

Motherwell's contract involved numerous variations.

Motherwell – Subcontractors/suppliers

Motherwell subcontracted out various parts of its work that it felt could not be done in-house. Cable trenching was one example. Other subcontractors included design consultants. To do the work, Motherwell needed eight full-time design engineers. This was not possible from in-house, and so Motherwell hired four design engineers. This had the advantage of having bodies to use when needed. The disadvantage was having to control the people who had no experience in the Motherwell quality system.

Motherwell used purchase orders to buy its supplies. This process was not done centrally but by individual foremen for all equipment except the major instruments, pressure transducers, digital indicators etc. This led to an increased cost over utilising a central buying scheme and having a central store for all items like relays, conduit, cable ties etc.

EXERCISE
1. Suggest an alternative delivery method for the project, given that:
   - There were many construction variations, and Homee only had the ability to approve minor variations.
   - Mehmet had little input to the problems experienced by the contractors.
   - Mehmet had to perform the overall commissioning of work done in parcels by the contractors.
2. In what aspects is your delivery method better? In what aspects is it worse?

4.9.3 CASE STUDY – OWNER-BUILDER

This case study discusses the issue of owner-builders, from a consumer's point of view. It highlights a number of facts overlooked by most owner-builders in their search to save money, particularly in regard to good contracts management practice.

Owner-builders are usually exempt from compliance with builders registration legislation, provided they are constructing their own house, and it is not for immediate sale. There may be requirements for insurance and compliance with local council guidelines, as well as redress against poor workmanship.

Background

For 18 months, Mr X worked on his dream of building his own home, a dream that is shared by many. Many people think, like Mr X, that they have enough time and ability to employ trade contractors, do some or most of the work themselves, and save a lot of money by cutting out the builder. The idea looks tempting and easy.

After completing a standard form, Mr X was issued an owner-builder permit. He started to shop around looking for the best quality material for the best possible price and, in the mean time, calling for quotations from trade contractors to do the work. The criteria for selecting the contractors were evidence of a licence and the cheapest offer, believing, like many owner-builders, that being a licensed tradesman is enough guarantee to have the work done properly and on time. And, because of the same belief, Mr X didn't give much attention to having a written agreement with every contractor, detailing the work, the cost, the timeframe and a penalty provision in case of delay or bad performance. Mr X thought that a verbal agreement was enough to form the relationship between the owner and a contractor.

As the work started, so the problems also started. The nice and friendly tradespersons started to show different faces. Each started to shunt part of the agreed trades work onto somebody else to do, and this ended up with Mr X paying extra money for almost everybody to finish their work as originally agreed. By the time the house was completed, the planned time had doubled, the proposed cost had escalated, Mr X had to lodge two claims to the relevant government authority for unfinished work by contractors, and the dream became a nightmare.

EXERCISE

1. Although it is true that owners, who intend to renovate, undertake a dwelling extension, or build a new home, can sometimes save money by coordinating and supervising the work themselves, this is not all. Many unknowns, disregarded by home owners, can easily upset the owner's time and money calculations. Things like the cost of holding charges, that is the cost of rates, land taxes, interest on money outlaid on the project before or during construction etc could turn a money saving proposal into a loss maker. Is the key to the owner's success in achieving this goal, good planning and a good contracts management practice, or are there wider issues?

2. From Mr X's experience, the cheapest offer from a contractor to do work was not necessarily the most economical one In many cases it was used by contractors to entice Mr X into a deal, with a lot of hidden extras to be claimed later on. Or it resulted in winning more jobs than the contractors could ever handle, and this resulted in delays to Mr X. In other cases it reflected bad workmanship and low quality finished work, or the intention of the contractor to use the cheapest possible material. That could be driven by the contractor's belief that owner-builders lack the experience or are not competent to detect their mistakes. Is choice of contractors on price alone the culprit here, or would such things as quality issues arise irrespective of the criterion used to select the contractors?

3. Many owner-builders expect more work from the trade contractors than they have described, when asking for a quotation, and others change their mind after the work has started.

   Disputes between owners and contractors are likely to arise when not enough care has been taken to describe the work, any requirements regarding quality of materials to be used by the contractor, the quality of the finished work, the cost of the main work and any future variations and a timeframe for starting and finishing the work.

   Is the owner-builder in a position to do all this competently?

4. Would engaging a builder, and hence transferring from a construction management type arrangement to the traditional arrangement, eliminate all the troubles Mr X had?

4.10 CONCESSIONAL METHODS

BOOT, BO, BOT, and BOO as well as BOLT, BLO and similar are acronyms referring to similar delivery methods, typically where financing is outsourced in addition to other parts of a project. (B = build, O = own, O = operate, T = transfer, L = lease) They commonly (though not necessarily) apply to private sector involvement in public sector projects and facilities; the private sector invests in facilities to be used by the community. They can also refer to private sector projects and facilities. The investor is typically a consortium of organisations with interests in financing, operating, and contracting.

They alternatively may be called *concessional methods of delivery* or *commercial development* strategies.

Numerous agreements may be necessary to get such developments going – *heads of agreement* include leases, financing agreements, guarantees, ..., while there are also the usual project contracts.

The concessional type arrangements are not recent phenomena, though in recent years they have attracted considerable attention. Railways, airports, power stations, canals, ports and telecommunications, for example, are areas in which private investors/promoters in the past have obtained *concessions* from a government to operate the facilities after having invested in their development. Some current applications include tunnels for vehicles, power generation facilities, water treatment facilities, prisons, hospitals, casinos, hospital car parks and roads (tollways). The concession period (the period of private sector involvement in the project/facility) is different with each case, but typically might be 20 to 40 years.

Design and construction of the facility may follow any of the available delivery methods, but design-and-construct with a fixed price is common. This may be carried out by a contractor separate to the investor-consortium, or the contractor may be one of the parties to the consortium.

While operating the facilities, the investors receive a return on their money. The client may agree to pay based on production or output. For example, an investor in a water treatment plant may be paid per kilolitre of water treated and supplied to a standard, with possibly a minimum guaranteed volume per day being paid for by the client. An investor in a gaol or hospital may be paid per person-day that the inmate or patient, respectively, occupies the facility, with possibly a minimum guaranteed person-days per year being paid for by the client. An investor in a casino may get its return from the gamblers, with a proportion of the turnover going to the government, as well as rent going to the government. Toll roads may operate on a profit sharing arrangement.

The key to whether a concessional project gets off the ground is its 'bankability'. That is, investors have to be convinced that they will get a return on their money, and that the associated risk is acceptably low.

With ownership over the concession period, the investor may, if approved, do additional things with the facility. For example a tollway investor may also operate service stations/fast food outlets adjacent to the roadway. With operation rights only, without ownership, such additional revenue generators may not be possible.

At the end of the concession period, ownership transfers to the client.

Examples

A correctional centre was procured based on a build and operate (for a set period) contract. One of the tenderers was in-house – the government correctional services department.

A hospital car park was procured on a build, lease and operate basis. Tenderers were car park operators.

A coal mine engaged a company to design, build and operate a coal washing facility on the mine site. The company is paid per tonne of coal washed, in return for financing and operating the coal washing facility.

Sydney Harbour Tunnel – a road tunnel. Sydney Harbour Casino.

Contract documents

The contract documents in a BOO(T) project usually include:
- *Concession Agreement* – between the government agency and the ownership vehicle.
- *Shareholder Agreement* – between investors.
- *Site Lease* – between the government and the ownership vehicle.
- *Design Agreement* – between the ownership vehicle and the design consultants.
- *Construction Contract* – between the ownership vehicle and the construction company.
- *Equipment Supply Agreement(s)* – with supplier(s).

- *Credit Agreement* – between the lenders and the ownership vehicle.
- *Lenders' first securities* – (over the infrastructure and ownership vehicle's rights under all contracts/agreements).
- *Enforcement Arrangements Deed* – between the government agency and the lenders.
- *Offtake Agreement* – the agreement whereby the 'product' of the infrastructure is sold.

(Clayton Utz, 1995)

Stakeholders

Example stakeholders in concessional type arrangements are shown in Figure 4.29, though many other configurations are possible. In some cases, the same organisation may fill more than one role.

Sometimes concessional schemes are drawn like Figure 4.30. Other entities are involved, for example material and equipment suppliers, insurers etc, but the key players are indicated in this figure.

Usage

There are a number of reasons why the public sector (and also the private sector) might adopt concessional type arrangements. These include:
- The public sector may be restricted in its borrowing limits. This restriction might be a government-imposed limit, or one suggested by credit rating bodies. Borrowing limits are a means of managing the public sector debt. Infrastructure that is desired, but which would require funds beyond any limit, could be procured through private sector financ-

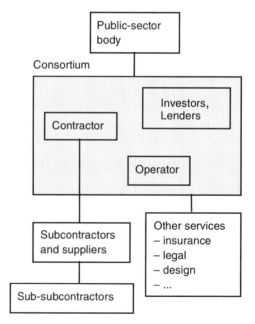

Figure 4.29. Example stakeholders in a concessional type scheme.

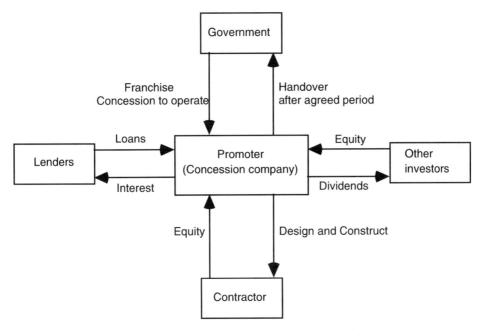

Figure 4.30. Example structure of a concessional scheme.

ing. Uncertainties may arise, when considering particular project proposals, as to whether they fall within or outside the imposed borrowing limits. Excessive borrowing can affect a government's credit rating.

Any proposal for private sector involvement may have to be couched in a format that doesn't lead to any public sector accounting liabilities, such as residual guarantees or other revenue effects.

- The inherent uncertainty of the budgeting process of government affects the predictability and timing of the delivery of important infrastructure. Funding is budgeted on a year-to-year basis and changing priorities over the years in which a project is underway can lead to it being delayed, adversely changed and occasionally abandoned. The private sector has a capacity to get on with and deliver an end-product, particularly when driven by business imperatives such as saving on interest costs.
- In most countries there are heavy demands on existing infrastructure, and a need for upgrading or new infrastructure. This requires a large amount of capital and technical expertise.
- Where the proposal comes from outside the country, it leads to direct foreign investment.
- Private sector involvement is said to lead to efficiencies through competition. The completed facility is said to cost less than if delivered by conventional means. Some people may hold a different view to this. As well it is very difficult to demonstrate.
- The private sector already has an involvement in the development of infrastructure, through, typically, design and construction. Concessional arrangements take this involvement a step further.

- Financing institutions may regard the private sector as being able to guarantee delivery of a facility in a shorter timeframe than the public sector. This certainty of outcome on the project completion may be appealing. The certainty of the contract, whereby it may be difficult to change, may also be appealing to financing institutions. The private sector's ability to operate a facility efficiently is also a consideration.
- Private sector ingenuity is said to be better than public sector ingenuity. Some people may hold a different view to this. The proposition that the private sector can add value to the delivery of public infrastructure is well accepted within government. Requests for contributions range from design-and-construct alternative proposals through to the design, construct, finance and operate options.
- The introduction of the private sector into the operating elements of infrastructure provides an opportunity to effect workplace reform and also introduce efficiencies of operations dictated by business rather than political considerations.

Characteristics

Characteristics of concessional type arrangements include:
- Relatively high (front end) costs are associated with preparing and promoting proposals. This may not always be appreciated by the public sector body. This can be a deterrent to many in the private sector participating in concessional type arrangements. The cost of arranging shortlists and proposals, although significant, are much less. Should the proposal be unsuccessful, then the costs of preparing and promoting proposals are thrown away unless some contribution has been made by the public sector body towards development costs.
- A user pays principle. The payment, however, may not reflect the true cost or benefit to the community. This is a controversial issue because the public at large is used to the idea that infrastructure should be provided by the government. It is submitted that as long as there is an alternative available to the users, infrastructure developments aimed at facilitating economical development will eventually benefit the community at large.
- The particular combination of financing, designing, constructing, operating, maintaining and owning chosen, will depend on a case-by-case analysis of each project. It will generally follow from an analysis of the needs and goals of the public sector, adjusted to take into account potential private sector involvement.
- Time periods of private sector involvement in a concessional scheme will vary depending on the desires of the particular private sector bodies. Designers, delivery managers, finance brokers etc may only wish a short term involvement. Longer term involvement will come from those responsible for the operation and maintenance of the facility.
- The lead role in any consortium proposal may vary depending on the project/facility. The lead may come from construction contractors (for projects involving significant construction), utility suppliers (e.g. power stations), equipment/plant suppliers etc.
- Financiers may only have an involvement ending at the end of the project work; they are looking to have an early return on capital. Alternatively, financiers may have a long term involvement, and an interest in the created financial market.
- Return on invested funds will depend on the risk profile of the project/facility and whether the equity provider wishes a short term or long term recoupment.

- The requirements of financiers, and the provision of finance, can be crucial to the success or failure of any venture, as is the availability of funds.

Some idiosyncrasies and problems

- The public sector body may not understand the needs and goals of the private sector participants, and vice versa.
- The argument is often advanced that the facility obtained through private sector involvement will cost less than if delivered by conventional means. Private sector finance, however, is usually more expensive.
- Tax advantages, real estate and so on might be offered to the private sector in order to render the development of the facility or asset feasible.
- Public sector bodies have changing priorities, while budgets are on a year-to-year basis. This may lead to potential projects being changed, abandoned or delayed.
- Once a contract is entered into, it may be difficult to take an alternative course, should the public sector wish to.
- Probity issues in public sector dealings with the private sector may dictate how proposals may be initiated and dealt with. This may cause frustration with private sector participants.
- There is a need to balance registrations of interest, short listing and proposal preparation.
- There is a trade-off required between the public sector body prescribing precisely the facility requirements and operating and franchise conditions, and the private sector being given sufficient flexibility to come up with a workable and perhaps innovative solution.
- The preservation of intellectual property from any innovative proposal is important.
- There is a trade-off between a risk allocation desired by the public sector body and that considered acceptable by the private sector.
- Government processes, such as approvals, need to be carefully integrated with the proposal development.
- Negotiations and initial leg-work in developing proposals may be lengthy and hence costly in terms of peoples' time.
- Financiers may decide not to commit themselves to any particular proposal until the successful proposal is known.
- Consortia membership, the selection of the best members, and agreements between members need sorting out early on in any development work.
- Good communication is required between consortium members, and other stakeholders.
- Financiers may force unacceptably large risks on contractors by requiring high levels of security guarantees and bonds. The contractor should only be asked to carry a risk that it can manage.

Risk sources and risk

Some possible sources of risk (for various stakeholders) include:
- Losing the tender after investing in a proposal; project not proceeding.

- Agreements between consortium members.
- Time, cost and quality matters during delivery; timely completion.
- Operation of the facility.
- User demand for the completed facility, product, consumable.
- Financier delivery and operating requirements.
- Changing legal and political environments.
- Insolvency of any of the stakeholders.

Some of the risk may be transferred to others by way of insurance. Success seems to come from managing the project/facility risks.

Commonly, whether a concessional project progresses depends on how the promoter is able to reallocate risks, offer guarantees, reduce uncertainties and so on.

4.10.1 CASE STUDY – WATER TREATMENT PROJECT/PLANT

Background

The water authority responsible for the supply of water to the community called, firstly, for expressions of interest, and then tenders for the design, build, own and operation of the water treatment plant. The own and operation period was 25 years.

The authority's reasons for using a concessional approach were:

- To use world's best practice in the design of water treatment plants resulting in reduced cost.
- To externally finance the project.
- To outsource the project, i.e. have a smaller water authority.

A consortium consisting of a developer, investor and a water treatment specialist company won the contract.

The consortium subcontracted the project management and design and construction to a civil engineering contractor (a subsidiary company of the developer company) (see Fig. 4.31).

Plant and project description

The water treatment plant takes water from a reservoir and purifies it to a certain standard for which the water authority pays an agreed tariff depending on such things as the water demand and water quality at supply and discharge.

The water drawn off the reservoir is treated with a flocculating agent, before it is passed through sand filter beds, and then chemically adjusted for acidity, taste and colour. The water is stored in temporary storage or balancing tanks. The filters are cleaned by backwashing and the solids removed by centrifuge.

The water treatment specialist company brought world's best practice in the design of water treatment facilities to the consortium. This reduced the cost of the project, and helped this consortium win the contract over competitors.

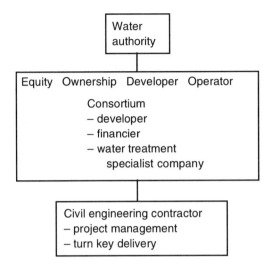

Figure 4.31. Contractual arrangements, water treatment project/plant.

Partnering

The water authority introduced the idea of partnering, whereby all parties would work together in a co-operative atmosphere of good faith and fair dealing. This was in addition to the contract. The civil engineering contractor used a similar relationship in dealing with its subcontractors.

Subcontracts

The civil engineering contractor decided to manage the project by subcontracting the work in three major packages: earthworks, concrete and mechanical. During the tender process, subcontractors were selected for their expertise to assist in the costing and design development, and on the understanding that if the contractor was awarded the contract, that the subcontractor's tender may be negotiated with the subcontractor. The subcontractors were bound to exclusivity and confidentiality agreements. The idea was that all parties were to work together with a common goal of winning the project – hence the idea of a partnering agreement.

After winning the contract, design proceeded to a point where a lump sum tender was negotiated with the subcontractors. If a tender could not be negotiated, open tenders were called.

The advantages to the contractor were:
• A learning curve experience for the contractor.
• Able to get subcontract cover and still have confidentiality of bid.
• If unable to negotiate a lump sum with the selected subcontractor, the contractor still had the option to call for other tenders.
• Contractor was able to tap subcontractor expertise.

The disadvantages to the contractor were:
- Less control over the performance of the subcontractor.
- With fewer subcontractors, there was more reliance on the performance of the subcontractors. Careful selection was required of the subcontractors, and whether they could work together.

The advantages to the subcontractors were:
- There was the opportunity to negotiate their tenders rather than being involved in 'hard nose' tendering.
- More control of their own work.

EXERCISE
1. The cost of preparing a tender for a concessional project and promoting that tender is very high, often in the millions of dollars. The assessment of the various bids is extremely involved, requiring skills that the assessment team may not have. There may be too much focus on the price, and not on the comparison of the quality or merit of the bids. There is much to support the idea of payment of costs of unsuccessful bid teams and this cost added to the total cost, as is sometimes done in design-and-construct projects. What is your view?
2. The competition to win such projects is commonly high, often to the point of 'buying' projects. In submitting a bid, a tenderer looks for a competitive difference or edge. In the above project, the water treatment specialist brought a better and cheaper product to the consortium's tender. How else might a competitive edge in tendering be gained in concessional projects?

4.10.2 CASE STUDY – HOUSING VILLAGE

A developer-consortium of two firms was contracted to a government authority to provide housing (equivalent to a village) for several thousand residents, after temporarily being used to provide accommodation for athletes and team officials for a major sporting event. This followed an open call for expressions of interest, and subsequent requests for further submissions for the planning, design, construction and financing of the village.

The various project agreements were:
- A project development agreement between the government authority and the consortium – development rights for the land, and government contribution, in return for providing temporary accommodation for athletes and officials. The government authority negotiated the project development agreement with two shortlisted bidders prior to announcing the winner. This enabled a rapid commencement of the project.
- A joint venture agreement to establish the roles and responsibilities of the consortium partners. One party was responsible for development, finance and construction delivery, and the other for the design, marketing and sale of the product.
- A finance facility agreement between the consortium and financiers to provide debt funding – project-based funding backed by appropriate corporate guarantees; interest paid by the consortium in return for financing the development.
- A delivery services agreement between the consortium and a contractor to act as the agent to procure design and construction services. A prime cost, fixed fee contract. The contractor acted as an agent for the developer, and when directed by the developer entered into contracts with consultants, trade contractors and suppliers on behalf of the developer.

- Fixed fee paid to the contractor in return for management services related to design, construction management services, reporting and cost control.
- As the consortium's agent, the contractor entered into professional services agreements with consultants to provide design services, and trade contracts to procure construction trade works.
  Consultants – fixed price contracts (lump sum or schedule of rates) and some prime cost contracts.
  Trade contractors, suppliers – fixed price contracts (lump sum or schedule of rates), some prime cost contracts and purchase orders for supply-only contracts. There were three different contracts – a major works contract and two types of minor works contracts – depending on the value.

The above series of agreements were designed to distribute risks to the parties most appropriate to manage the particular risks.

Delivery method

The delivery method for the project could be described as Build, Lease and Sale (BLS) because the consortium-developer never became the owner. Instead, the title remained with the government authority until transferred to the end purchasers of the houses, with roads and parklands being under management of the local council. Financing was outsourced in addition to other parts of the project.

The delivery method for the project, including all contractual links is shown in Figure 4.32.

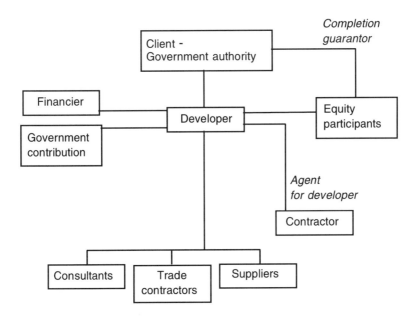

Figure 4.32. Village project delivery method.

The chosen delivery method was used to:
- Draw on private sector skills and expertise.
- Minimise government funding (budget limitations).
- Enable private sector funding.
- Quarantine the government from managing industrial risks.
- Obtain innovative solutions to minimise the sales risk from developing dwellings to suit the sporting event requirements, in lieu of being produced at a rate, which could be absorbed by market demand.
- Minimise project development cost due to private sector competition.

The characteristics of the method were:
- A relatively high cost associated with preparing the final bid submission (two remaining competitors). This was partially offset by the government reimbursing partial bid costs to the unsuccessful bidder.
- The developer's high risks, due to the market uncertainty. There remained an uncertain time period for the developer's involvement.

EXERCISE
1. What is your view on the suitability of the delivery method and contractual links for such a project?
2. The financing institutions regarded the developer as being able to guarantee delivery and the outcome of the project in a shorter time frame than the government. Would this be a general truism of private sector versus public sector, or only a perception?
3. The contractor acted as an agent for the developer instead of as a head contractor. This eliminated the contractor's liability for any costs, damages, and expenses suffered by the developer; the contractor also did not assume liability for the performance of consultants or trade contractors. In such circumstances, how does the developer ensure that the contractor is acting in the developer's best interests?

4.11 FAST-TRACK

In some projects it is desirable or unavoidable that the project phases overlap. A project is said to be fast-tracked if its phases overlap (Fig. 4.33).

At the activity level, a similar situation occurs when start-to-start or finish-to-finish links are specified on the activity network, in place of more usual finish-to-start links. This has the effect of completing the work sooner, which is the intent of fast-tracking.

A common occurrence of the fast-track approach is where the design and construction phases overlap; that is, the design is incomplete before construction starts (see for example, Fig. 4.34).

The term 'integrated construction management' may be used to refer to the case where the design phase and construction phase are integrated or overlapped. This, in part, refers to the construction management delivery method being more amenable to fast-tracking than the traditional delivery method.

Side effects of fast-tracking may be:
- Procurement requirements for the implementation phases are uncertain.
- Unnecessary implementation costs with implementation based on design assumptions.
- Inefficient design.

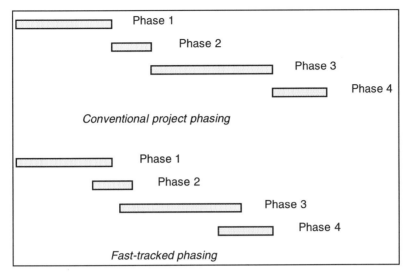

Figure 4.33. Bar chart showing comparison of conventional and fast-track approaches.

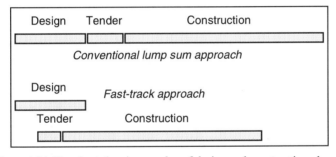

Figure 4.34. Bar chart showing overlap of design and construction phases.

- With staged or progressive approval from relevant government authorities and the owner, there is a risk associated with subsequent stages not being approved in a preferred form, and having to accept something less than a first choice.
- Extra management effort and capabilities are required over non fast-tracked projects.

- *Reliance on cost estimates to indicate the limit of the [owner's] financial commitment until completion of the works.*
- *Under fast-tracking, the [owner's] prerogative to make changes to the design is greatly curtailed.*
- *The up-front cost of the construction phase is usually higher but it should be off-set by the buildability of the design and the shorter overall project development period.*
- *Much greater exposure by the [owner] to financial risk.*

(Uher & Davenport, 1998)

Example

As a trade-off between earlier completion and extra cost, footings for a building may be over-designed, excavated and poured before the superstructure is designed.

Benefits of shorter project time

- Facility available earlier; less holding charges.
- Return on income sooner.
- Less rise and fall costs.
- Transfer from existing facility sooner; rental avoidance on existing facility.
- Less risk of change of user requirements.

Costs of shorter project time

- Extra cost to speed up program.
- Extra control needed.
- Less thorough planning.
- Project progress may not match capabilities and resources.

4.11.1 CASE STUDY – INFRASTRUCTURE PROJECT

The case study discusses the construction in a major civil infrastructure project for a government agency by a (part) design-and-construct – (part) novation contract. This multi-discipline project comprised earthworks, stormwater drainage, retaining walls, bridges, tunnels, tunnel electrical and mechanical services, roadworks and utility adjustments. The engineering design for the project was complex, given the necessary integration of a range of engineering disciplines and the owner's requirements for high quality and timely delivery of the project within a limited budget.

The key aspects of the project were the contractual arrangements and the program.

Contractual arrangements

The contractual and administrative arrangements for the project were involved and required considerable co-ordination and interfacing between the various parties throughout all stages of the project (Fig. 4.35). The key parties included the owner, the owner's project manager, the owner's consultants, the contractor and the contractor's design consultants. Furthermore, the owner was accountable throughout the project to a number of other government bodies which formed a project control group. A formal partnering approach was adopted on the project by the owner and a number of major partnering sessions were held during the course of the project.

Pre-contract award phase

The project was separated into two major components defined as Subproject -Area 1 and

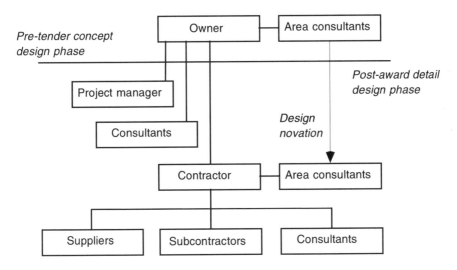

Figure 4.35. Contractual arrangements on infrastructure project.

Subproject-Area 2. In the pre-contract award phase of the project, the owner engaged a number of engineering consultants for Area 1 to prepare a preliminary concept design, which satisfied the owner's requirements and was able to be used as a basis for design-and-construct tender proposals. The owner also engaged an architectural consultant for Area 2 to prepare a preliminary concept design for the purposes of novating the design for this part of the work to the contractor.

All contractors bidding for the project engaged a major engineering consultant during the tender phase to assist in quantifying the scope of the works in Area 1 to be detailed designed and constructed during the post-contract award phase.

Post contract award phase

The owner engaged a project manager and a number of other consultants to assist in the delivery in the post-contract award phase. Following award of the contract, the successful contractor engaged their pre-contract award consultant to undertake the detailed design and documentation of the work as principal design consultant for Area 1. The principal design consultant, in turn, engaged the services of a number of sub-consultants to assist in completing all components of the detailed design work. The contractor also engaged a number of subcontractors and suppliers and a number of secondary consultants for some specialised aspects of the project.

Approximately three months into the contract, both the preliminary concept design of Area 2 and the owner's architectural consultant were novated to the contractor.

Project program

The project included a very tight work program, which resulted in a fast-track design and construction process that ensured the contractor was able to achieve early construction milestones and the owner's overall contract completion date (Fig. 4.36). Fast-tracking resulted in the design and construction phases overlapping, and hence construction starting before the design was complete.

The pre-contract award phase activities were compressed into an extremely short timeframe primarily due to political pressures. The post-contract award phase activities were also compressed into a very short timeframe. Early construction works began within a month of detailed design commencement of Area 1. Pre-novation design for Area 2 was undertaken at the same time as the detailed design and construction of Area 1 were being carried out. The construction of Area 2 commenced approximately midway through the construction of Area 1.

Procurement

The delivery method adopted by the owner was a combination of partial lump sum design-and-construct, project management, design novation, partnering, and fast-tracking as well as direct owner procurement for some supply items. The delivery method adopted by the owner was chosen to meet the requirements of both the project and the owner. These requirements were timeliness, value for money, suitability and quality. In arriving at the preferred method the owner considered the following constraints:

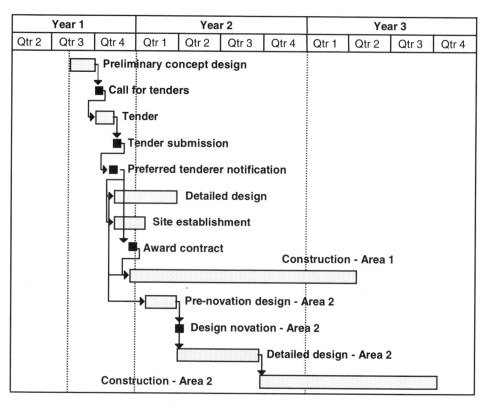

Figure 4.36. Program for infrastructure project.

- Significant time constraints and political pressures due to a fixed completion date.
- Pressures to work within a government approved upper limit project budget.
- Significant and complex site constraints created by existing features and the future construction of facilities, including the need to interface with adjacent construction contracts.
- Provision of a facility of high quality (incorporating 'best practice') to satisfy current and future stakeholders in terms of design compliance, aesthetics, environmental factors, operation, maintainability, safety etc.

Traditional delivery was considered better suited where time was not as critical and where the optimum design for the project could be confidently predicted without the expertise of an experienced contractor. Design-construct-and-maintain delivery was not suitable for this project because the owner was intending to lease the operation and maintenance of the infrastructure for an extended period of time. Commissioning and handover phases of the project were omitted because these phases required the input of a number of government authorities and this was considered to be more appropriately managed by the owner. A concessional type contract was not considered in detail as finance for the project had already been pre-arranged and was immediately available from government funds. Construction management was not considered relevant to this type of project.

Preliminary concept design

The final form of design-and-construct contract (i.e. 'part' design or 'full' design) required careful consideration by the owner.

The owner wanted to develop a very clear and reasonably comprehensive design-and-construct brief for this project to ensure that all the various stakeholders' requirements and site constraints were fully identified. A preliminary concept design, including a combination of well-defined constraints and performance type specifications, was developed for Area 1 by design consultants engaged by the owner, to form the basis of the design-and-construct contract.

The owner was therefore able to have some control over the direction of the final design, whilst still enabling design and construction innovation/optimisation by tenderers, possibly leading to a more efficient and cost effective project for the owner. Furthermore, the owner was transferring the significant risk associated with the design and the design-construction co-ordination to the contractor.

Design novation

The owner also wanted to have considerable control over the design of Area 2. Following the award of the main contract, the design of Area 2 was developed by the owner's architectural consultant to '40% status', after which time the architectural consultant was novated to the contractor to complete the design and documentation under fast-track conditions. The owner was again transferring the risk associated with the design (and possibly also the pre-novation design) and the design-construction co-ordination to the contractor, although the owner was likely to be paying a higher premium to the contractor for this novation method due to the lack of competition.

Project management

In addition to the design-and-construct contract, the owner awarded a separate project management contract to an engineering consultant to manage the design and construction work on this project (and adjacent projects), and to provide technical advice and contract administration services.

The owner considered the engagement of a project manager to be essential due to the size and complexity of the project, the need to undertake extensive co-ordination and management of interfacing contracts, and the need for ongoing review and refinement of the project scope to ensure the owner's requirements for the project were being met. The owner was also involved in the day-to-day running of the project and overall cost management, and was a government body without the necessary experience to carry out all of these management duties.

Partnering

Due to the complex nature of the project and the large number of interested parties, the owner selected a formal partnering arrangement for the project in an attempt to get all parties together to work towards agreed project goals and to improve co-operation and communication.

EXERCISE
1. Due to the perceived complexity of the design and construction for the project and the necessity to fast-track the project in order for it to be completed on time, the owner believed that the above delivery method was the most appropriate form of delivery method for this project. How else might the project have been procured?
2. *Advantages to the owner* – The primary advantage to the owner of using design-and-construct on this project was that the delivery method was able to handle a complex and difficult project within a very short time period and at a competitive price. It is unlikely that all of these critical project requirements could have been completely fulfilled by any other method of delivery.

   Another key advantage of the design-and-construct method for the owner was that it allowed design-and-construct teams tendering for the works to be innovative, and this could result in more efficient and economical solutions for the project. Considerable design effort was required and the cost of this effort was borne by the contractors bidding for the project. No tender compensation payments were made by the owner.

   During the tender period for this project, one contractor developed a technical solution for a portion of the work which was significantly different to the solution proposed in the owner's preliminary concept design. The alternative solution, however, complied fully with the owner's brief and offered considerable benefits to the owner in terms of reduced disruption to facilities during the construction phase. This solution had not been addressed adequately by the owner's consultants in the development of the preliminary concept design.

   The design novation of Area 2 was reasonably successful from the owner's perspective in that it allowed the owner some considerable control over the design direction whilst still leaving the risk of the design and the construction with the contractor.

   Suggest other advantages to the owner.
3. *Disadvantages to the owner* – The contractual arrangements and fast-track nature of the project however did create a number of problems along the way for the owner.

   The owner was not able to control or contribute significantly to the design development of Area 1 as much as intended, primarily due to the fast-track process and the involved contractual arrangements which allowed little design development and review by the owner. For example, the design and documentation for a portion of the work was submitted by the design consultant to the contractor, then passed onto the owner's project manager, following review by the contractor, then onto the owner following review by the project manager, and sometimes then onto the project control group following review by the owner. Any comments the owner (or project control group) may have had often could not be implemented, because in the meantime either the design had advanced significantly or the construction of that portion of the work had commenced.

The owner was also not able to achieve as high a quality of the constructed work as was envisaged, primarily for the reasons outlined above and also because the owner relied too heavily on the contractor's 'quality system'. In former projects, the owner may have had a site representative or site team to continuously monitor and supervise the contractor's activities to ensure that the construction works were being carried out in accordance with the design documentation and contract requirements. For this project, the owner's project manager did not perform this role in a full time capacity and similarly the design consultant was required only to undertake periodic surveillance.

The owner also discovered that there was less flexibility to make changes to the scope of the contractor's work at minimum cost due to the fast-track process. A number of changes had been required due to the impacts of other interfacing contracts under the control of the owner. Often there was little room for the owner to negotiate time extensions or costs associated with some variations due to the threat of delays to the contractor's program. The supply of incorrect information or slow response to critical requests for design information by the owner or owner's project manager also resulted in delay costs which would otherwise not have occurred in a more conventional contract arrangement. The management and co-ordination of the contractor's design, and response to the contractor's queries was a very difficult task for the owner and the owner's project manager.

Suggest other disadvantages to the owner.

4. *Advantages to the contractor* – The design-and-construct method had a number of clear advantages for the contractor.

During the tender period, the contractor was able to explore alternative options which best suited the contractor's plant and equipment or were simply cheaper and better solutions not identified in the owner's concept design. On this project, the innovative solution proposed for a portion of the works contributed significantly to the contractor winning the contract. On the other hand, the short tender period placed a significantly higher level of risk on both the contractor and the consultant in the tender design and pricing of the works, than would be the case under the traditional method.

Post-contract award, the contractor was able to control the design process more effectively by directly engaging the consultant and thereby influencing the design to suit the contractor's constructability requirements, whilst still satisfying the owner's brief. The contractor was also able to deal with a number of design issues without necessarily involving the owner or the owner's project manager. Closer interaction between the designer and the contractor enabled considerable cost savings in some areas of the project to be realised by the contractor.

The fast-track nature of the project was also generally good for the contractor, because of reduced overheads and administrative costs.

Suggest other advantages to the contractor.

5. *Disadvantages to the contractor* – The contractor was reliant on the design consultant producing fast and accurate designs for early procurement and construction purposes, and this was not always possible. The design consultant was often restricted by slow or inaccurate responses to requests for information forwarded up the hierarchy to the contractor, owner's project manager, owner or other third parties. Furthermore, the design consultant was under considerably more time and resources pressure than on a more conventional contract and this could have led to mis-

takes and subsequent rework due to inadequate checking, lack of design co-ordination and/or incorrect assumptions where design information was lacking. Fast and accurate responses, to design queries, by the owner and/or owner's project manager is essential in a fast-track design-and-construction process, but was lacking on this project in a number of areas.

Inefficient design can also appear in design-and-construct projects due to the lack of time available to adequately develop and/or optimise design and also in the often drawn-out approval process for alternative designs. Inefficient design can lead to additional costs overall if additional material and labour costs are not offset by actual construction time savings. On this project, for example, some design alternatives were proposed which deviated (with sound reasoning) from government standards. However timely approvals were not forthcoming either directly from the relevant government bodies or indirectly from the project control group. One of the problems encountered on this project was that the contractor wanted both design development (to optimise design and therefore reduce costs) and early final documentation (to shorten construction time) and this, in a number of areas, was simply not achievable.

The design novation of Area 2 was also frustrating for the contractor, in that the owner was still wanting significant input into the detailed design process, well after novation had taken place.

Suggest other disadvantages to the contractor.

6. *Partnering* – Partnering on this project from the contractor's perspective was not particularly effective, although it did enable representatives from all key parties to come together and meet each other and gain a better understanding of the goals and requirements of the various parties. Whether a formal partnering process was in place or not on this project mattered little to the contractor because the formal contractual process for most activities still needed to be followed. Although a significant number of informal meetings between the various parties were able to be held from time-to-time to resolve important problems between the parties, it was considered by the contractor that these meetings would have more than likely occurred anyway. It was the contractor's view that the success of the project still relied very heavily on the pro-active and co-operative nature of the individuals representing the various parties, and on this project this was evident in some areas and not in others. Comment on this as a general experience on other projects.

4.11.2 CASE STUDY – INTERNAL UPGRADE PROJECT

This particular production-line project was not approved (by senior management) until four months after the date that the project group had estimated was required in order to complete the work to a tight, but conventional program. To maintain the end date, the project schedule was partially crashed. The justification for the same end date was that the project needed to be completed by then to suit funding and functional requirements. Another consideration was that the earlier the project was finished, the greater the financial return to the company.

The project commenced with the basic design minimally complete, due to restrictions in spending and the inability to place fixed contracts. Many contracts were started with a letter of intent; several companies did not commence any detailed design work until their contact was signed. The problem with this was that the company still expected and required delivery by the commitment dates. The design was then crammed into a shorter period, and many errors were found during construction, perhaps due to the limited time given to check and cross check information on drawings. The design was still being progressed as construction commenced.

A major section valued at 25% of the project took a step change in direction three months following project approval. To meet the completion date, the approach to the task required considerable overlapping of activities, with several major operating/functional issues unresolved prior to construction commencing, let alone the design being completed. The project manager also reduced the aim date for completion of start-up to a month short of the required finish date. The reduction in duration came about through changes in direction and overlapping phases, with a major increase in project staff required to progress the completion of the design and the construction in parallel.

Altered methods of procurement were required to achieve the new project completion date. Extensive planning was required to achieve the final date without major incident. The delivery methods revolved around splitting the project into smaller work packages, that could be tendered as grouped work packages. This involved a further breakdown in activities in order to determine what sections could be done independently of other activities, and readjustment of priorities. The schedule was dramatically restructured to accommodate early starts for work packages that could be overlapped.

Due to a variety of other delays, activities were being managed on a priority list to suit the design, construction, commissioning or training requirements. The planning of activities was critical for activities to run in parallel. The number of personnel to complete the activities or track information was doubled, with design and drafting personnel added to the team to try and shorten the time to complete activities. The coordination of the design then became a process solely aimed at keeping up with construction. Cooperation between disciplines was paramount with negotiation and coordination between all personnel required. The failure of the team to accommodate these issues resulted in additional costs on site.

There were problems with people failing to recognise requirements due to a lack of communication on particular activities that were in progress. Many errors were made because of incomplete or rushed designs and the overlapping of activities. There was a cost penalty with major parts built with incomplete designs, or the design having to be changed to accommodate site completed activities. The result from this was a significant amount of rework to repair, modify and ensure the plant was suitable for continued operation. There was a significant cost penalty for the increased rate, or desire for project completion.

The construction contractor for one section of the project was brought in early to try and streamline the design and construction process. The method used to establish the contract was through negotiation with a previously known contractor. The result was a preliminary costing that exceeded the original estimate expectation. The senior project engineer decided to adopt the design changes recommended by the construction contractor. The major change in the design approach was done without significant thought placed on the downstream costs and time. The issue was examined in isolation and failed to note the downstream effects on other work packages.

The statement made at the time by the senior project engineer was, 'do not complicate the issue with time, this is a matter of money'. The design of the structure had to be completely reworked to accommodate the different construction approach. The extreme overlap of the construction contractor commencing work and the changes to the design created large inefficiencies in the design process, with the design team being unable to keep up with the construction program. The subsequent result was the company incurring considerable costs from construction delays and design changes, as well as ongoing difficulties in the downstream functions. On reflection, the change in design cost more than it originally was intended to save, both in time and money.

The ongoing work to complete the project then dealt with the compromises made in the early construction work to complete a project to a fixed date.

EXERCISE
1. Is it true to say that in fast-track projects, you have to expect that there will be things that no one has thought of beforehand, and there will be extra costs? Or can fast-track projects be budgeted reasonably tightly?
2. The completion of the project early had considerable benefits for the company, making it desirable to fast-track the project. The partial crashing of the project schedule had already been done pre-project approval (because the start had been delayed) in an attempt to match the original completion date for the project. The risks associated with fast-tracking this particular project were considered acceptable. In other situations it may be unavoidable to overlap the project phases and so have a situation develop that is considered fast-track. What are the risks associated with fast-tracking, and what makes them 'acceptable'?
3. In fast-tracking, the procurement requirements for the implementation phase are uncertain. Delivery methods are best selected to suit the situation as it occurs during the project. The preferred company method of competitive tendering was compromised, in order to progress the procurement process quickly. There were occasions of uncertainty as to the best approach for minimising the time to complete. The preferred approach was to break the sections of the work into like areas and have multiple contracts that allowed a greater overlap in activities, rather than negotiate with particular contractors, though this was done when required due to time constraints. What other approaches are available for minimising the time to complete?
4. The decision to dramatically change the design, because the initial negotiated costs far exceeded expectation, led to a snowballing effect where the step-change caused more costs than originally anticipated. Can the consequences of cost be considered in relative isolation to time and a fixed completion date, as occurred in this instance? If a step-change in design is made to reduce the identified, construction costs, without examining the ongoing costs associated with redesign and significant delays that may occur, are anticipated savings lost before they are realised?
5. In part of the project, the overlap between design and implementation was very pronounced, with the consequence being an inefficient design and a loss of time due to this inefficiency. Do fast-track projects allow people to be realistic in this regard? For example: the initial saving of time may be lost in the large percentage of rework that is required to correct the errors made because the design was incomplete before issue. The desire to start early must be realistically examined with regard to the percentage of design that is complete. What saving there is in changing the design may also be effectively lost in the cost to redesign the structure and the delay caused by the change in direction.
6. The pressure on people to reduce time and save money through compressing schedules, in hindsight in this situation, cost more than could be saved by overlapping the design and construction phases to the extent that occurred. How do you ensure enough allowance is made in the overlap, to minimise construction delays due to delays in design information or material procurement (if occurring separately)? These delays cost time and consequently money. Would a lower risk approach in this instance have been to have the contractor do the detailed design and more of the material procurement?
7. Unnecessary implementation costs are frustrating. Are they more likely to occur when the design is being developed in conjunction with construction, because design changes incur greater costs when the structure is built and modified rather than changed on paper? How should the trade-off between additional costs, that occur when the design is incorrect, against the 'time lost', by having the design complete prior to implementation, be considered?

8. Is it the case that, if the project manager and team wish to aggravate stress levels and spend additional amounts of money and time on coordinating and organising, then fast-track a project? Is it necessary for the project manager and team to be certain of their ability to keep on top of the issues that develop during a fast-track project, else severe cost penalties occur with inefficiencies?

9. 'Fast-track is a coordination nightmare, a planner's hell, a designer's heartache and a site construction manager's death wish'. 'Time is money'. Is it the case that, if you get it right it is great, if you get it wrong it is an expensive mistake?

10. 'No matter how well the plan is developed, some events, that cause havoc, happen.' Is this a truism for fast-track projects? Some of these events are when the owners change the project's constraints. Reducing the end date, extending the start date, and changing the design mid-project are all activities that can cause havoc with a well thought plan. The project, as described, had a shortened delivery date, an extended start time and a major design change shortly after approval, all changes that had considerable effect on the time schedule and the project plan. But should these necessarily lead to havoc?

11. 'The retention of the quality in a project is problematical when the fast-track approach is used'. In some areas of the project, quality was compromised to achieve the end date and considerable amounts of rework were required for long term functionality. This was a concern that was noted by the project team in some areas but ignored on occasions in favour of reducing time. Does quality necessarily have to suffer in fast-track projects?

12. 'Fast-tracking might be viewed as the administration of multiple projects, for a single project, that results in the overlapping of the various operations'. How useful is such an observation?

13. Fast-track success requires a high degree of planning, knowledge of the operation involved, a flexible but decisive mental approach, very close coordination, direction and control, effective cooperation and teamwork between all project parties. When determining the contractual and organisational arrangements for the project, how might you take these considerations into account?

14. The approach taken by the project changed during the implementation stage, causing considerable changes in the project team. The team was not originally established to accommodate a fast-track approach. This made the process more difficult, with progressive changes in team dynamics occurring at later stages in the project than would ordinarily be acceptable.

There is considerable debate in the project management literature and in the practice of project management about the appropriateness of fast-track. Strong views are held on both sides of the argument. Fast-track is not seen by some as a cheaper overall option and it may not even save time. Express your views.

4.12 SINGLE OR MULTIPLE CONTRACTS

The provision of work, services or materials may be divided up (in a number of ways), if it is thought there is some advantage in multiple contracts (*packaging*) as opposed to a single contract. This may be referred to as a *divided contract approach*. Alternatively, in the other direction, aggregation reduces the number of contracts (see for example, Fig. 4.37).

The choice of delivery method may determine whether single or multiple contracts are used in any situation. For example, the construction management delivery method, by its very nature, packages the work into multiple trade contracts.

The type of project, owner requirements etc may determine the use of single or multiple contracts. For example, project financing may require a staged approach. Projects may

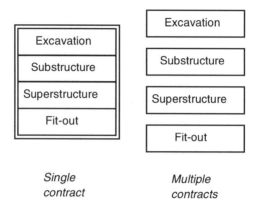

Figure 4.37. Example single or multiple contracts.

have natural division points, e.g. substructure, superstructure and fit-out of a building; piers, superstructure, approach roads and landscaping for a bridge. The substructure work may contain uncertainties and hence the type of contract used may be different to that for the more definite superstructure work. Mechanical and electrical work may be separated.

4.12.1 BUNDLING

In a project of any size there is commonly the option to let the work as one large contract, or multiple smaller contracts, or something in between. Where work parcels are combined to give a single contract in place of several, it may be referred to as *bundling*.

Bundling of contracts is a useful way of reducing contract administration resources. It can also provide economies of scale if done properly.

Example

An energy supplier, rather than having twenty contracts for different types of cable, has one supplier for all twenty types. With one supplier, tendering time is reduced, while the relationship between customer and supplier can be developed. Similar bundling is used for stationery products.

Single contract

Advantages
- Reduced need for the owner to coordinate separate contract packages.
- Single point of responsibility for administration, and legal pursuit.

Disadvantages
- Reduced flexibility to respond to owner or contractor changed status.
- Requires a well-defined project.

Multiple contracts

Advantages
- Contracts may be for sequential and/or concurrent work; work-based or trade-based.
- Allows contracts to be awarded as documentation becomes ready; fewer variations because contracts not let with incomplete documentation; less need for provisional sums.
- Allows staging, for example in response to funding or cashflow constraints; assists expenditure control.
- Allows early start.
- Allows fast-tracking, and project compression; allows program acceleration or deceleration.
- Allows the project to change direction; standards can be reduced, work omitted to assist budget; standards can be increased, work added if funds and time become available; new or fast-changing technology can be incorporated.
- Enables direct hire of specialist consultants and contractors, and direct purchase of materials/equipment; assists quality control of the finished product.
- Suitable for work that is physically fragmented, for example by distance; suitable for a project with disparate subprojects; suitable for isolating high risk aspects (e.g. foundations).

Disadvantages
- More control, than by a single contract, is required by the owner.
- Increased supervision by owner; increased fee to consultant/agent if present; increased administration if work in contracts is simultaneous.
- Possible delay claims caused by actions of other contractors; liquidated damages affected by actions of other contractors.
- Multiple separate or independent contracts may be suboptimal, e.g. design may not consider constructability.
- Coordination and interface risks carried by the owner.
- Coordination issues; possible bickering between contractors; possible lack of cooperation between contractors; possible one 'weak sister' contractor affects performance of other contractors.
- May encourage owner to make changes because of the flexibility of multiple contracts.
- Need to ensure that the extent of the work is completely covered; uncertain division of responsibility.
- Need to ensure consistency or no discrepancies between contracts.
- Firm price not known till all contracts have been let.
- Increased record keeping.

Issue matrix

The New South Wales Public Works Department (1993) gives the matrix of Figure 4.38 to assist in the assessment of risks for the single and multiple contract cases. The figure lists the major issues that might affect an owner, and assists in the selection of the most appropriate delivery method.

To use Figure 4.38, those issues that are considered critical (carry the largest weighting

[Issue]	[Issues regarded] as critical	Single contract	Multi- contracts
Cost and/or time impact due to coordination difficulties		L	H
Inability to control cashflow		H	L
Cost impact if redesign during [execution*]		H	L
Cost impact if project curtailed during [execution]		H	L
Inability to economically fast-track to achieve early commencement of completion		H/M	L
Variability of end cost to pre-[execution] budget		L	H
Inability to react to technological changes economically		H	L
Cost and/or time impact of contractor's failure to complete contact(s)		H	L
Cost and/or time impact due to documentation errors between contracts		L	M
Cost and/or time impact of individual documentation errors		H	L
Inability to directly select subcontractors/suppliers		M	L
Inability to economically amend/impose changes in staging		H	L

Figure 4.38. Choosing a delivery method (New South Wales Public Works Department, 1993). *execution = manufacture, fabrication, construction, ... ; H = high; M = medium; L = low.

for that project) to the owner are identified, and then the single versus multiple contract options are examined against these issues. In either case there is a need to consider the risk analysis further for the detail within individual contracts.

4.12.2 STAGING

For a number of projects over time, either in the same location or of similar type, the owner may elect to procure the work in stages or sequentially, with a series of contracts. The terms *staged*, *serial* or *sequential* may be used in such cases. The same contractor may or may not do the whole work, but the one desire on the owner's part is to achieve consistency in finished product. Tendering is carried out for each stage. The contractor who wins the first stage may have some advantage over competitors for the second and subsequent stages, in terms of reduced establishment costs, an existing working relationship with the owner, and inside knowledge of the owner's requirements. Accordingly bids may be low for the first stage, in order to win the subsequent stages.

Fast-tracking may be associated with the practice of using multiple contracts. For example, as part design work is completed, a contract for its construction is awarded; this is without waiting for the full design to be completed.

Examples

Ongoing health works programs, and school and tertiary education building construction may be handled on a stage-wise basis. There may be several school buildings on the one site, or school buildings in different localities.

Continuation contracts
A continuation contract differs from a serial contract in that it is an ad hoc arrangement to take advantage of an existing situation. Thus, if a housing contract is already going ahead on a particular site, a further site may become available with similar housing requirements. In this case there will have been no standing offer to do more work, the original tendering documents having been conceived only for one particular project. However, they can provide a good basis for a continuation contract.

Alternatively, it is possible to make provision for continuation contracts in the tendering documents for the original project, and this is sometimes done. But there is no commitment and the possibility of a continuation contract may not arise.

(The Aqua Group, 1975)

4.13 PERFORMANCE OF DIFFERENT DELIVERY METHODS

Much blame is attributed to the traditional method of delivery for cost and time overruns of projects; hence the search and experimentation with other delivery methods. However, no objective evidence appears to have been published which can directly attribute a project's woes to using the traditional method. Instead, it is more likely that the project's woes have been the result of the inability of project participants to work together as a team, managerial inadequacies, inadequate planning and control, unrealistic estimates, poor information and communication systems etc.

Projects, being one off, prevent there being any definitive conclusions being obtained as to the fault or no-fault of the delivery method used. There are so many other variables on projects that affect the project outcome. It is not possible to do 'double blind' or unbiased experiments. No rigorous scientific-style experiment is possible.

A number of authors have compared different delivery methods and their influence on cost and time overruns. Such reported studies tend not to be conclusive for a number of reasons, including:
• The era and locality in which the studies were carried out affects the outcome.
• There are multiple uses and misuses of the terms describing the different delivery methods, and without specific details in the reports, the exact delivery method is unknown.
• The influence of different payment types is not considered.
• The influence of different conditions of contract is not considered.
• Benchmarks for what are reasonable project durations and project costs are unknown

relative to the tendered durations and prices, and hence the significance of cost and time overruns is unknown.

- Were contingencies included?
- Was rise and fall included?
- Was the project cost/contract price and duration adjusted along the way?
- Was the project subject to competitive tendering or by negotiation?
- Were similar management practices, similar skills and similar expertise adopted between projects?
- Size and type of project affects the outcome.
- Truthfulness in reporting project outcomes – projects are reported as being successful even though the reality may be different; project failures and inadequacies of project personnel tend to be covered up.
- Confidentiality of actual project outcomes, particularly financial.
- Different risks being taken by the project participants on different projects.
- The extent of changes in the design during construction.

The point to note from all this is to be wary of someone who says that one particular delivery method has been found to be better than another. There is no conclusive evidence for this. All that can be said is that someone 'feels' that one particular delivery method is better than another. Without hard evidence, this gives rise to *fashion* in the use of delivery methods, and a 'gravy train' for contract consultants and legal advisers, playing on the indefiniteness. As well, the contract consultants and legal advisers cannot be accused of recommending the wrong delivery method to clients, because 'wrong' essentially can't be proven. Fashion in the use of delivery methods can be seen in the recycling of the popularity of the different delivery methods, typically under new names; most delivery methods have existed for many years, few new ideas are put forward but new names do arise for old ideas. Few people take the time to look at historical practices and past publications to realise that the wheel is being, not so much reinvented but rather, renamed continuously.

The situation is one of 'horses for courses'. That is, examine the project situation, and choose the most appropriate delivery method. Don't blindly use any delivery method.

Articles frequently appear in industry magazines expounding the virtues of one particular delivery method, because perhaps it has worked well on several of the (article) author's recent projects. The views are always subjective and unjustifiable in the general context, and no rational appraisal is given. Folklore or a tide of popularity may follow regarding a particular delivery method, until some other authors publish criticism of the particular delivery method, because perhaps it hasn't worked well on several of these other author's recent projects. The views are always subjective and unjustifiable in the general context, and no rational appraisal is given. Users then search for the next panacea. And so the process goes on, with the state-of-the art not advancing.

Risk, duties and obligations

Each of the delivery methods represents different duties and obligations, and a different sharing of risks, between the project participants.

For example, in management contracts, the owner has greater involvement and assumes a larger financial risk. Design consultants become designers only, removing any lead or managerial role that they might have in other delivery methods. The contractor is expected to be proficient in management methods, and is paid a fee. The contractor assumes little risk unless a guaranteed maximum price contract is used. Agency-type agreements provide better security of payment for specialist contractors, and may be preferred because of this. Under a guaranteed maximum price contract, specialist contractors may be subject to bid shopping.

Fragmentation

The project industry appears to becoming more fragmented, with project participants becoming more specialised. This places greater emphasis on coordinating the project participants and managing the interdependence. Whether specialisation in the industry is desirable or not is uncertain. Certainly, getting a team atmosphere on projects is more difficult with specialisation. Against this, specialisation offers efficiencies and skills unattainable through generalists.

With fragmentation, has come the trend to using delivery methods that capitalise on fragmentation. This may have long term implications for training and employment of workers in project industries, and this in turn may see a return to less fragmentation and larger organisations in the future.

There is no guarantee that any delivery method will produce the desired outcomes. The reason is that the delivery method, that is selected, provides only the structural framework within which actions are taken and decisions are reached. It is necessary to understand how that method operates in practice by examining behaviour within the system, and the external factors influencing this, as well as looking at the contract documents that specify formal duties, rights, and obligations. The external factors may include the relationships and the personalities of all parties involved in the project. The combination of all these factors will produce a particular set of outcomes.

4.14 EXERCISES

Exercise 1
a) Figure 4.7 lists a number of criteria, for which ratings are asked. What is it about the design of the ratings scales that enables the selection chart (Fig. 4.8) to be drawn?
b) Can the different delivery methods be bundled into three groups, or is this too simplistic?
c) What improvements would you suggest to the approach for selecting a preferred delivery method?

Exercise 2
Compare what you perceive are the main advantages and disadvantages of detail design-and-construction over design-and-construction.

Exercise 3
Compare traditional and design-and-construct forms of delivery methods according to the characteristics in the following table. Do the analysis from the owner's viewpoint. Use entries in the table such as: low/medium/high; small/medium /large; etc.

Characteristic	Traditional	Design-and-construct
Length of tender period		
Cost impact of design changes		
Cost and time impact of coordination difficulties in design and construction		
Impact of design changes on original contract price		
Cost impact of variations due to documentation errors		
Potential for lesser design quality		
Potential for inclusion of contractors' contingencies		
Cost impact of latent conditions		
Cost impact of other variations		
Potential for time overrun		
Ability to fast-track		
Potential for contractor's claims		
Time till contract award		

Exercise 4
a) What do you see as the disadvantages of the owner nominating subcontractors and suppliers to be used by the contractor?
b) Is novation a means of achieving the advantages of nomination, without the associated disadvantages?

Exercise 5
List what you perceive are the advantages and disadvantages of a project management consultant also performing a role as one of the consultants, for example a design consultant, and assuming that the consultant has some technical expertise.

Exercise 6
Two projects are reported below from the trade contractor's or subcontractor's viewpoint.

a) On *the first project,* the owner engaged an independent consultant to prepare, and be responsible for, the concept design, detail design and the contract documentation including the specification, and a construction management organisation to organise the construction, and manage the project.

The construction management organisation was responsible for producing the construction program and the control of resources required to maintain production to reach the milestones and deadlines highlighted in the construction program. It was also responsible for contract administration and the co-ordination of trades and services.

The construction management organisation was used as a 'letterbox' by the trade contractors, with regard to requests for information, approval of shop drawings, claims for variations etc.

All project correspondence for the aforementioned items had to pass through the construction manager, who would then forward it onto the appropriate body, i.e. owner, architect or the design consultants, who in return would reply with an appropriate answer or instruction passed through the construction manager's office back to the contractors or suppliers, who had initiated the request for information, shop drawing approval etc. (see Fig. E6a.)

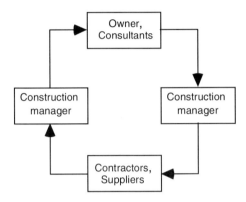

Figure E6a. Project correspondence loop of particular construction management project.

As the construction manager was not responsible for the detail design drawings or specifications, any delay in replies to requests for information, approvals of shop drawings, variations or changes to the design documentation was used to extend the construction program. The construction manager seemed to have an indifferent attitude to any delays or changes caused by the owner or the design consultants, because it was not responsible for those delays. This, in turn, led to increased costs and time, and disputes.

How might you address such issues within the construction management or project management frameworks?

The contractors were committed to the construction program and the production levels to maintain it. It was the owner's responsibility to manage the interface between the design consultants and the construction management. In this instance, the owner built a wall between the design team and the construction team, due to its lack of experience in the construction industry, and this led to coordination and constructability problems.

b) On a *second similar project,* the design-and-construct method was used. The owner engaged an independent consultant to prepare the concept design brief, and a contractor to prepare the detail design and documentation as well as manage the construction and commissioning within a lump sum tendered price. The contractor, in turn, engaged consultants to complete the design and subcontractors to do the work.

All requests for information, approvals of shop drawings and claims for variations by the subcontractors were handled quickly and efficiently by the consultants and architect engaged by the contractor. Any delay in approvals of shop drawings or replies to requests for information, that may have had an impact on the construction program, were highlighted and pursued by the contractor, so as to minimise any delays to its construction program. Also, any changes to the detail design drawings or specifications, other than those required by the owner and which may have resulted in additional cost, had to be substantiated by the consultants or architects to the contractor, because these would have to be paid for from its fixed price (see Fig. E6b.)

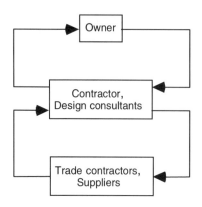

Figure E6b. Project correspondence loop of particular design-and-construct project.

The subcontractors were committed to the construction program and the production levels to achieve its milestones and completion dates and, due to the fact that the design team and the construction team worked with each other with minimal formality, this project progressed smoothly, with very little increase in time or costs, and therefore few disputes. The problems associated with possible detail design and approval delays were those of the contractor.

Based on these two projects, there is an obvious reported bias towards design-and-construct over construction management, at least from the trade contractor's or subcontractor's viewpoint. Is this a fair comparison between the two methods? What is missing from the comparison?

REFERENCES

Antill, J.M. 1970. *Civil Engineering Management*, Angus and Robertson.
Australian Constructors Association (ACA) 1999. *Relationship Contracting*, ACA, North Sydney.
Barrie, D.S. & Paulson, B.C. 1992. *Professional Construction Management*, McGraw-Hill.
Clough, R.H. 1960. *Construction Contracting*, Wiley.
New South Wales Public Works Department (1993), Capital Project Procurement Manual, Construction Policy Steering Committee, New South Wales.
Clayton Utz 1995. Mini-Boots – Private Funding of Smaller Infrastructure Projects, *Australian Construction Law Newsletter*, Issue 40: 37-38.
Davenport, P. 1993. Pitfalls in Novation, *Australian Construction Law Newsletter*, Issue 29: 38-42.
Department of Defence 1992. User Guide [to delivery methods contracts], Canberra.
Eves, C.G. 1995. Recent Overseas Experience in Structures for Private Roads, *Australian Construction Law Newsletter*, Issue 40: 39-47.
Institution of Engineers, Australia 1995. Re-Engineering Competitiveness Inquiry, *Issues Paper*.
NPWC/NBCC 1990. *No Dispute*, National Public Works Conference, National Building and Construction Council, Canberra.
The Aqua Group 1975. Which Builder?, Granada.
Uher, T.E. & Davenport, P. 1998. *Fundamentals of Building Contract Management*, UNSW.

BIBLIOGRAPHY

Design-and-construct
Davenport, P. 1991. Design and Construct Contracts, *Australian Construction Law Newsletter*, Issue 16: 30-32.

Pilley, J.L. 1996. AS4300-1995 Design and Construct Contract Published, *Building and Construction Law*, 12: 2-6.

Tyrril, J. 1995. Standards Australia's Design and Construct Contract AS4300-1995, *Australian Construction Law Newsletter*, Issue 16: 17-18.

Design novation

Australia Is Using New Type of Contract 1990. *Engineering News Record*, Vol. 225: 44.

Project management

RAIA 1979. Project Management, Practice Note PN63, RAIA.

Construction management

Tyrril, J. 1989. Project and Construction Management Agreement, *Australian Construction Law Newsletter*, Issue 2: 18-20.

RAIA 1983. Construction Management, Practice Note PN77, RAIA.

Concessional

Jones, D.J. 1994. Key Issues for Major Bot Projects, presented to the School of Civil Engineering, The University of New South Wales, 7th April.

Shaw, G.& Haley, G. 1992. Infrastructure Development, *Asian Architect and Contractor*, February, pp. 42-46.

Tiong, L.K.R. 1990a. Comparative Study of BOT Projects, *Journal of Management in Engineering*, *ASCE*, 6(1): 107-122.

Tiong, L.K.R. 1990b. BOT projects: Risks and Securities, *Construction Management and Economics*, 8: 315-328.

Tiong, L.K.R. 1995. Risks and Guarantees in BOT Tender, *Journal of Construction Engineering and Management, ASCE*, pp 183- 188.

Tiong, L.K.R. & McCarthy, S.C. 1992. Critical Success Factors in Winning BOT Contracts, *Journal of Construction Engineering and Management, ASCE*, pp 217-228.

Tyrril, J. 1990. NSW Private Sector Infrastructure Provision Guidelines, *Australian Construction Law Newsletter*, Issue 14: 43.

Fast-track

Kerzner, H. 1997. *Project Management*, Van Nostrand Reinhold, 6th ed.

Lock, D. 1997. *Project Management*, Gower, 6th ed.

Young, E. 1994. 'Fast tracking' in Lock, D. (ed), *Handbook of project management*, Gower Technical Press Ltd.

CHAPTER 5

International case studies

5.1 OUTLINE

A number of contract practices from different countries are presented here. Many of the practices, as will be observed, are not necessarily peculiar to the country of the case study. The intent of the case studies is to illustrate a range of practices.

5.2 CASE STUDIES

5.2.1 Case study – Port development, Asia

Ports P/L has a significant investment in concessional development work. Of 25 sites that are currently managed by Ports P/L, or Ports P/L has investment in, over 90% are of this type of contractual arrangement.

The basic investment criteria used by the Ports P/L in establishing new joint ventures worldwide are:
- Identification of key port locations.
- Large capital input with 60:40 gearing.
- Ports P/L equity ideally 40% with management control.
- Non-recourse finance.
- Maximum two year construction period for 'greenfield' site.
- Graduated profit margins during the phase-in period.
- Management fees to provide operational and technical expertise.

The first step in the investment process is to identify a potential facility. In some instances, the company is sought out and requested to submit a proposal for investment in a port facility. Whatever the origin, the company conducts a feasibility study to satisfy itself as to the potential of the project. Once the company is satisfied that a port is feasible, it then submits a comprehensive proposal to the appropriate authority.

From the point of approval onwards, it is a matter of mobilising the personnel who will start the project and in most cases ultimately manage the completed facility. This covers operations, financial management, human resources management, information technology and engineering services. Often this work has to be achieved within a few months of the concession being obtained, and Ports P/L total resources provide the necessary capacity to achieve such tight deadlines.

One container terminal

A consortium/joint venture was established between Ports P/L and a local company. Ports' investment was about 95% of the project, and was sourced from the parent company, and a loan from the World Bank.

The licence was awarded to the consortium to build ($200M), operate and manage a 600 m container terminal for a period of 30 years.

The return on investment for the terminal was based on a long-term model, and hence the 30-year concession that was secured. The terminal was a greenfield site and had strong competition from established terminals.

Developing the port saw the reclamation of 20 acres, with the wharf and approach bridge requiring 800 steel-lined piles and 21,000 m^3 of reinforced concrete.

Structure

The structure of the companies forming the consortium is shown in Figure 5.1.

Generally, the financing, designing, constructing, operating, maintaining and owning aspects are similar in all of the projects that Ports P/L becomes involved in, but does vary in terms of the source of finance as required by the local authorities, and their preferences for type of investment.

In the structure of the consortium, the lead role was taken by Ports P/L as the major investor and the supplier of the management resources. The project management company was a Ports P/L company and the design consultants had been used by Ports P/L on a large number of its greenfield sites, but was an external organisation.

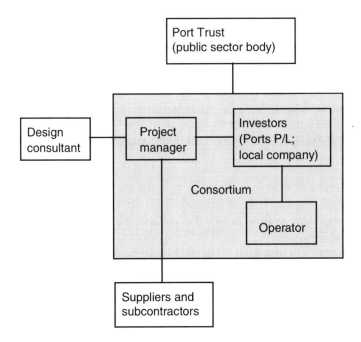

Figure 5.1. Consortium arrangements for the port development.

A number of set-up teams were deployed to the site to implement: the Terminal Control System, Finance Management Information System, and the Computerised Maintenance Management System, which are a suite of applications Ports P/L have developed and integrated for use in greenfield projects.

The types of companies used as suppliers and subcontractors were: civil and building construction, essential services, port equipment manufacturers, and watercraft manufacturers.

Usage

The local government was firmly committed to increasing privatisation of its infrastructure, be it ports or roads, telecommunications or urban services. A series of tax incentives and other concessions were announced and regulations and procedures considerably simplified. Most investment proposals were, in fact, automatically approved.

The local government had recognised the benefits of inviting investment and stated:

Recognising the need for major expansion of port infrastructure to handle increased foreign and coastal trade, the government has thrown open the sector to private participation. The goals: to introduce competition in port services, improve efficiency, productivity and quality of service; reduce the gestation period for setting up new facilities; and bring in the latest technology and management techniques.

Risk

There were a number of risks associated with this project. The most significant being that this was the first port privatised by the local government and substantial negotiations were involved in securing the concession period length required to satisfy the investors. Concessions often given elsewhere may have been only half the period or less, but generally the capital investment would not be as large.

This facility was the first time Ports P/L had undertaken a joint venture in the region, but it was comforted with the knowledge that it had successfully taken over the management of and built up a facility in a neighbouring country. The local government had indicated that there were many more of its major ports that would be available for privatisation and Ports P/L was keen to perform well in this market to be invited to tender on these future ports.

Incentives

For any investing company to be involved in concessional type work with public entities, incentives are necessary.

The local government promoted investment in the ports sector through:

Approval incentives
- Automatic approval of equity up to 51% in providing supporting services, such as operation and maintenance of piers, loading and discharging of vehicles.
- Automatic approval of equity up to 74% in construction and maintenance of ports and harbours.

- The government would consider proposals for 100% privately-owned holding/subsidiary companies for port infrastructure.

Taxation incentives
- Financiers – Keeping in view the large investments and long gestation periods that characterise infrastructure projects, various tax concessions were available to institutions engaged in their financing. Forty percent of the profits that long-term financing institutions made through infrastructure financing was tax-exempt, subject to specified conditions. Venture capital funds investing in power or telecommunications projects were exempted from taxes on dividend income and long-term capital gains. Infrastructure capital funds did not have to pay income tax on dividend or interest income, or long-term capital gains.
- Operators – A five-year tax holiday followed by a deduction of 30% for the next five years was provided to BOOT/BOT projects in the following areas: power, roads, highways, bridges, airports, ports, rail systems, water supply, telecom sector, irrigation, sanitation and sewerage systems.

These were all supported by the necessary legislation.

Tendering

Country guidelines for private investment in major ports were:
- Open tenders invited for private sector participation on a BOT basis.
- Bids invited based on a two-envelope (or two-cover) method, consisting of technical and financial bids. Only financial bids of technically qualified bidders opened.
- Evaluation of the bids based on the maximum realisation to the port by the Net Present Value method using a discount rate as periodically fixed by the government.
- Maximum licence period (including construction period) not to exceed 30 years.
- At the end of the BOT period, all assets to revert to the port, free of cost.

Tenderers to provide the following information:
- Up front fee for the lease or licence.
- Royalty per tonne of cargo handled.
- Minimum cargo which it would be willing to guarantee or pay as lease rent per unit area of land/waterfront.

Specific guidelines for port-based industries included:
- No financial return was being guaranteed to developers.
- The investor was required to keep any port property that was transferred to it insured. It cannot sub-lease, sell, subcontract or in any other way transfer any asset without the port's prior approval. It cannot abandon services abruptly or dispose of land, machinery or other assets or convert them even partially to non-port uses.
- The developer must comply with all the port-related statutes including labour laws.
- Environmental and other clearances for projects to be obtained by the port, or the investor, depending on the requirements.
- If the port prepares feasibility reports for the project, it will recover the costs from the successful tenderer.

EXERCISE

1. These incentives and initiatives overcome many of the barriers that usually exist in undertaking BOOT operations. What other main barriers, do you see, to a consortium entertaining the idea of a BOOT project?
2. What would be the most suitable way to structure the income for the developer in order to repay its investment – based on charges per ship? Fixed payment? Volume or weight of cargo? Other?

5.2.2 CASE STUDY – HYDROELECTRIC POWER STATION PROJECT, ASIA

Introduction

The Shang Hydroelectric Power Station project consisted of construction of a concrete dam, power house with 1400 MW generating capacity, and navigation lock. It involved approximately 4M m³ of earthworks and 3M m³ of concrete work.

The project was financed by international money. The contractors and suppliers in this project were chosen through international competitive tendering.

The owner of the project was the local power authority. The project was managed by a construction company (CC) specially set up for the project. The tender documents were prepared by a joint local American group.

The main engineering works were divided into five main contracts:

1. Civil works construction.
2. Turbines and generators supply.
3. Metal structure supply – heavy steel structures, related equipment and services.
4. Electrical and mechanical equipment supply.
5. Erection.

Each contractor had subcontractors or suppliers. There were also some small construction and supply contracts.

The contractor who undertook the civil works contract was called the major contractor on the project. The civil works contract was both lump sum and schedule rates. Items were detailed in a Bill of Quantities (BOQ). No adjustment to the price of the items under the lump sum component was permitted, whereas the schedule rates were adjustable subject to some conditions indicated in the conditions of contract.

It was the first time for many of the local participants to be involved in such a project which was conducted according to international construction practice. When the detail design was underway, the major construction contracts had been awarded. Thus, the design was restricted by the construction contracts. The design was expected to conform with the construction contract documents.

Review of the procurement

The approach employed in this project followed common international practice for procurement. There were few differences compared with common practice in the international construction industry. However, there were some particular considerations, when the tender packages were being prepared, where delivery methods that would best suit the conditions of the project were selected.

The procurement was divided into five major contracts, and five head contractors were

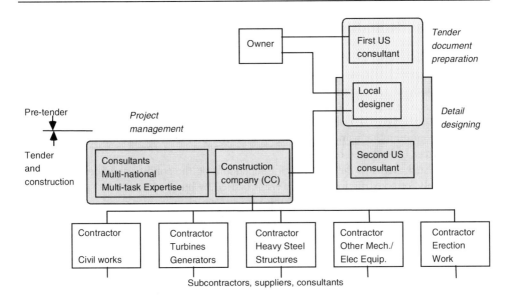

Figure 5.2. Contractual connections.

engaged. All of the main contracts were awarded and managed by the construction company (CC) that represented the owner, instead of the owner directly. Further, international expertise was widely used in this project. The contractual connections within the project are shown in Figure 5.2.

The advantages of this approach were as follows:

- From the owner's position, the whole project was completely under CC's control.
- Dividing a large job into several tasks, with appropriate purpose and size, could attract more interested tenderers. Otherwise, only a few potential tenderers might be eligible, and that would have led to less competitive bids.
- The approach shared the project risks among the several contractors.
- With several contracts, the budget control for both owner and individual contractors was easier.

The disadvantage of this approach was that the project management, particularly coordinating different contractors on such a large project, was difficult. The owner had to employ a fully experienced project management agency and additional expertise. The cost of project administration was high.

Civil works contract

The civil works were the largest component of the project, with the highest cost and longest duration. The type of contract was mixed lump sum and schedule of rates.

Most work under the contract was paid through a schedule of rates. The reasons for that were as follows.

- In such a large project, it was impractical to complete the whole detail design before the tenders were called. Also, when the contract work was underway, the electrical and mechanical equipment contracts had not yet been assembled. And so the exact engineering quantities were not known at that time. Changes to design were also expected. Considering these factors, a schedule of rates contract appeared the best way to minimise disputes.

• A schedule of rates contract avoids tenderers being put into a high risk position and hence raising the contract price to protect themselves. In addition, to further reduce the contractor's risk, the unit prices in the contract were designed as adjustable subject to some conditions, considering the long contract period. The method of adjustment of unit prices is discussed below.

However, such items in the contract not dependent on the conditions of design, or equipment supply, or construction period were procured by lump sum with no changes permitted. For example, all cofferdams of the project, which were designed in entirety, constructed and removed as a contractor responsibility, were paid for this way.

Civil works contract adjustments

The unit prices in the contract could be adjusted through either of two ways, as specified in the conditions of contract.

1. Each progress payment certificate in the foreign currency portion was adjusted by a Price Rise and Fall Formula. The Formula was designed to reflect the changes in the costs of salaries, materials, construction plant, and services bought with foreign currency by the contractor, as given below.

$$AF = 0.15 + 0.85\ (0.13\ *EP_i/EP_o + 0.10*S_i/S_o + 0.24*C_i/C_o + 0.03*T_i/T_o$$
$$+\ 0.45\ *P_i/P_o + 0.05\ *M_i/M_o)$$

where

AF = Adjustment Factor to the total foreign currency component of the progress payment certificate.

EP_o, S_o, C_o, T_o, P_o and M_o = the respective basic price indices for expatriate personnel, steel, cement, timber, construction plant and marine transport prevailing 45 days before the final date for submission of tenders.

EP_i, S_i, C_i, T_i, P_i and M_i = the respective current price indices for the same expenditure mentioned above ruling during the month before the month to which the progress payment certificate refers.

Price indices were those published at regular intervals by a government department or an officially recognised agency in the countries whose currencies were used as payment.

2. Adjustment of the unit prices of significant items in the Bill of Quantities might be made, if the actual quantity of such item varied more than 25% from the estimated quantity shown in the Bill of Quantities, at the time of award of the contract.

A significant item in the Bill of Quantities was defined as an item with an estimated amount of at least 5% of the total contract price, at the time of contract award.

For overrun, the first 125% of the quantity shown in the Bill of Quantities was paid for at the unit price quoted in the Bill of Quantities. The quantity in excess of 125% was subject to adjustment in favour of the owner.

For underrun, a total quantity of 75% of the quantity shown in the Bill of Quantities was paid for at the unit price quoted in the Bill of Quantities. If the actual quantity was less than 75% of the quantity shown, the unit price for the item was subject to adjustment in favour of the contractor.

The practice of contract price adjustment in the Shang project, as described above, achieved the result that the contractor was still bound by the price it submitted in the tender documents, while a limited price adjustment was allowed to partly realise the contractor's risk brought about by the long duration of the project. The adjustment was designed as being fair to both sides.

EXERCISE
1. Summarise the disadvantages and advantages of these procurement practices.
2. Comment on the logic of using five head contractors instead of one.
3. Comment on the fairness of the Rise and Fall formula.
4. Comment on the feasibility of completing the detailed design after contracts had been awarded, and the exposure of the contractor in this situation.
5. Comment on the role of foreign companies working within this delivery method.

5.2.3 CASE STUDY – HYDROELECTRIC PROJECT, ASIA

Introduction

This case study outlines the delivery method used in the Geng Hydroelectric Project.

The Geng project was a large project, by world standards. The completed facility was intended to control floods, generate hydroelectricity for an urban industrial centre, and to improve river navigation.

The major structures included a 185 m high gravity concrete dam, two group power-houses with a total 18,200 MW hydropower generating capacity, twin 5-stage locks, and a single 1-stage ship lifter.

The owner of this project was a specially established state corporation, which was set up to look after project development and project funding, and also operation after completion. The corporation established several functional departments: Construction Department, Plan & Contract Department, Technical Department, Accounting Department, Equipment Department, Material Department etc. to undertake construction management, budget management, technical management, and procurement management.

A local water authority entered into a fixed percentage fee contract with the owner to perform the feasibility study, conceptual design, item design, detail design, and preparation of project documents for tendering.

Supervision comprised nine consultant companies, and each also had a fixed percentage fee contract with the owner with responsibility for progress, quality, and cost control during construction (Fig. 5.3).

From the commencement of the project, there were more than ten main local contractors working on site. The contracts between the contractors and the owner were fixed price, mixing both lump sum and schedule of rates.

Permanent equipment and materials were ordered separately by the owner.

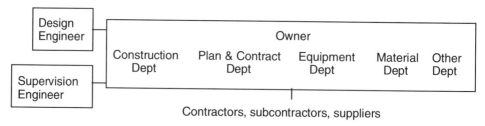

Figure 5.3. Responsibility for progress, quality, and cost control during construction.

Owner involvement

The delivery method was essentially the traditional method, but there was also some direct control. The owner employed many experienced in-house staff for construction at each stage. The huge amount of work was divided into separate main items, such as right bank excavation, left bank excavation, lock excavation etc. Contracts were issued for each separate item. The owner not only controlled construction progress and quality through supervising engineers, but also looked after design changes.

In-house work

The owner had its own Equipment and Material Departments to purchase permanent equipment, main materials and heavy construction equipment. The main materials were supplied to contractors according to an established plan or the contractor's material schedule. Heavy construction equipment belonging to the owner was leased to contractors to meet construction requirements.

There were five main reasons that influenced the adoption of the chosen delivery method:

1. As a large project, which was regarded as a symbol of achievement, the project had a large impact on the local community. And so the owner had to participate throughout in order to ensure construction quality and the completion of the project within established milestones and budget.
2. The project involved large amounts of work and required a large amount of capital. It was difficult to find a competent general local contractor. Importantly, the owner wouldn't diminish its control by introducing a general contractor.
3. The conceptual design and a large part of the detail design had been completed before construction commenced. It was possible for the owner to give clear requirements to the contractor and to issue contracts for separate items at different construction stages, according to an established schedule.
4. The owner employed many experienced technical people and construction managers who provided a sound basis for construction management.
5. After the project was completed, the owner intended to undertake other similar projects. Therefore, the owner could use its human resources and equipment in the long term.

The main advantages of the delivery method were:
- The owner could use outside resources and retain higher level control.
- The owner could gain an advantage by introducing competition.

The main disadvantages of the delivery method were:
- The owner faced a high level of risk.
- Tendering for each of the separate items caused higher tendering expenditure.
- Tendering and contracting for each separate item hindered the contractor's ability to make consistent long term working arrangements.
- Excessive sections and processes involved in control caused higher overheads and sometimes less efficiency.

Comparison of the Geng project and the Yau project delivery methods

The Yau project was located nearby and funded internationally. The main structures completed under this project included a 245 m high double arch concrete dam, an underground powerhouse with a 3000 MW hydro-power generating capacity, a tunnel spillway and a log pass facility. The owner of this project was a specially established corporation (Fig. 5.4).

Project comparison

Both delivery methods were largely traditional, but each had its own characteristics (see Table 5.1):
- The Yau project had less owner involvement. Heavy construction equipment and materials were purchased by the contractor directly. The contractor had high level control. Because the contractors were dominated by overseas construction companies, disputes often arose over different interpretations about contract terms.
- The Geng project involved more participants than Yau, but all participants were local companies. They had the same culture and norms, and communication was easy between each other. Sometimes, disputes arose from political issues. Therefore, although the delivery method in the Geng project was complicated, it still operated successfully.

EXERCISE
1. Weigh the advantages versus the disadvantages of the chosen delivery method for the Geng project.
2. Do the five main reasons given for adopting the chosen delivery method for the Geng project allow for other possible delivery methods? Discuss.
3. Can disputes be avoided over interpretation of contract terms when parties from different countries are involved? Discuss.

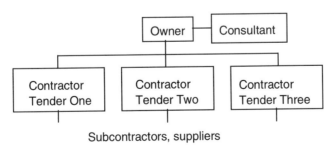

Figure 5.4. Delivery method in the Yau project.

Table 5.1. Project comparison between Yau and Geng.

	Yau project	Geng project
Owner	Government owned company; has professional staff.	Government owned company; has professional staff.
Consultant	Prime cost contract with owner to do both design and supervision.	Supervisor is different to designer.
Contractor	Three foreign-local joint ventures: – Tender one: civil construction, – Tender two: underground excavation, – Tender three: power plant installation.	More than 10 main locally owned construction companies; contracts for excavation, concrete, steel installation, ...
Contract	Schedule of rates contract with change rate for labour and material	Both lump sum and schedule of rate contracts.
Funding	International	National.
Design	Completed before construction	Substantially completed before construction.
Size of work	Large	Very large.
Job site	Located in narrow gorge, difficult to arrange construction facility. Avoiding construction interference was considered when planning the work.	Comparatively open area. Different contractors can work concurrently without much interference with each other.

5.2.4 CASE STUDY – CONDOMINIUM PROJECT, THAILAND

Introduction

The project owner was a real estate administrator, and had some experience in the construction industry. The owner wanted to control all the project phases, and also reserved the right to change the scope of works.

At the beginning, the owner engaged one consultant company – it included both architects and engineers of different disciplines in its company – to do the full design for the project.

After the design and documentation processes had been completed, the owner engaged a novice project management company to manage and control the project.

The delivery method for this project was similar to the project management method.

The owner let a contract with the consultant based upon a fixed price and the consultant was paid after finishing the design. The contract type for the project manager was the same as for the consultant but the owner paid a fee to the project manager every month for a period of 15 months. This was the period expected for the project. The project manager received almost the total amount of its fee.

For piling work, the owner, on advice from the project manager, engaged a contractor on a lump sum contract. After this work had been finished, the project was delayed for 10 months due to a shortage in the owner's finances. While in this delay period, the project manager calculated that there were not enough piles to support the building; at that time the piling contractor had left the site. The owner carried this extra cost due to an error of the consultants.

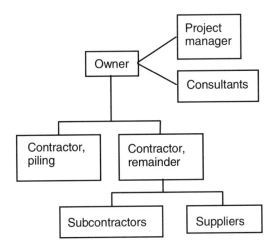

Figure 5.5. Delivery method used in the condominium project.

When the financial status of the owner recovered, the owner engaged only one contractor, by competitive bidding, to complete the rest of the work. If the contractor wanted to hire any subcontractors to do special work, e.g. electrical works, sanitary works, ..., the owner agreed.

A lump sum contract was let between the owner and the contractor. For variations, there was an agreement including a schedule of rates which provided a predetermined price for additional works if required.

The delivery method for the project is illustrated in Figure 5.5.

Comment on delivery method

The main parameters affecting the development of the delivery method were cost, time and risk allocation.

The duties of the project manager were only the control, management and coordination of the construction for the project, and as such the delivery method might be more appropriately called a construction management type.

This delivery method delivered several advantages. It helped the owner in refining the owner's requirements, which were anticipated to change in the construction phase. It also supported the needs of the owner through continual involvement in the project. The owner felt that an agent was needed to reduce risks.

The project manager was appointed after the consultants. Consequently, the project manager could not influence the design.

The delivery method was chosen to reduce the owner's contact to one contractor at a time, rather than, say, multiple trade contractors.

Comment on contract type

All contracts were lump sum contracts. For the consultant and the project manager, the

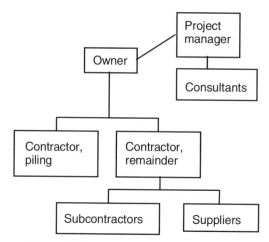

Figure 5.6. Alternative possible delivery method.

scope of the work was reasonably well defined and the documentation reasonably complete; therefore they could be confident of the return for effort, and the owner of the approximate total cost.

For the contractor (piling) and contractor (remainder), the scope of work and the documentation were clearly defined but they had to accept risks in order to complete the work within the lump sum; the owner preferred this option because the financial risks were transferred to the contractor.

The schedule of rates was added to the agreement in order to cope with variations. As one of the duties of the project manager, the quantities of added or deleted works were measured by the project manager.

In the market at the time, the owner used the lump sum contract as an incentive for the contractor.

EXERCISE
1. To improve the outcomes of the consultants and reduce the errors made by the consultants, the delivery method might have changed the link between the owner and the consultant to that in Figure 5.6. On the new link, the project manager would be responsible for the results from the consultant. The cost of the piling error would possibly not have been the owner's.
    Express your views on this opinion.
2. How is a lump sum contract an incentive, when the contractor is carrying all the risk?
3. What recourse is there to consultants for the piling design errors?

5.2.5 CASE STUDY – DEVELOPMENT OF PUBLIC FACILITIES, BUILDINGS, INDONESIA

Project description

The project involved the construction of several multi-storey buildings, infrastructure, public utilities, and the renovation of existing sport facilities for a national production company. The owner was a large, Indonesian company, which did not have experience in building construction.

The project was an important investment, and prestigious.

The project was divided into stages. This case study focuses on the first stage of the project – the construction of a festive market, sports club, infrastructure, and the renovation of an existing sports centre and hockey-ground.

The consultants

The owner hired two consultants to prepare the design. The first design team consisted of several overseas consultants and was responsible for developing the concept design and 85% of the detailed design for the entire work. The second design team consisted of local consultants and was responsible for developing the remaining 15% of the detailed design, in order to satisfy the owner's requirements and site conditions. This team worked on-site and assisted the project manager in dealing with alterations and variations as the work on site progressed.

The project manager

The owner engaged a project manager, whose role was to act on behalf of the owner in dealing with local consultants and contractors. The responsibilities of the project manager included the following:
- Managing and controlling project activities.
- Giving approval for the materials specification and shop drawings that were submitted by contractors.
- Giving approval for executing the project activities.
- Communicating the owner's requirements to the contractor.
- Reporting and evaluating progress work to the owner.

The contractors

The work covered quite a large area. Thus, the owner appointed two different contractors to develop the plan. Contractor 1 was responsible for the construction of the festive market, sports club, infrastructure, and the renovation of the existing sports centre. Contractor 2 was responsible for the construction of a parking-building and the renovation of the existing hockey-ground. All contracts were of the lump sum type.

As well as the large size of the project area, other reasons for appointing two different contactors were:
- To maintain quality – It was hoped that by employing two contractors it would create fair competition between the contractors to do their best work.
- To allocate the resources – Due to the complexity of the project, it would have been too risky if the owner had appointed only one contractor to do the whole work. The risks associated with labour shortages and lack of equipment might delay the project schedule.

See Figure 5.7 for the relationship between the project participants.

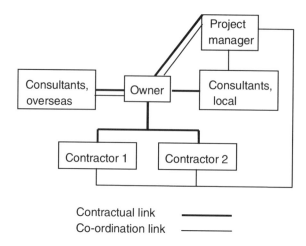

Figure 5.7. Delivery method, public facilities and building project.

Delivery option

The delivery method in this case used a management contract. The major factors influencing the delivery method were:
• Quantity and size of the work.
• Type of work.
• Resource availability.
• Technical complexity.
• Time, cost, and quality requirements.
The main reasons for the owner's decision to use the chosen delivery method were:
• The owner was from the production industry. It was not worthwhile for the owner to employ people to perform the project activities in-house. Moreover, the owner would not have similar projects in the future.
• The owner did not have experience in construction work. Although the volume of the work involved in this project was large enough, it was still inefficient for the owner to employ its own design-and-construction team.
The owner engaged two consultants to prepare both the concept design and the detailed design, while the project manager assisted in managing the interface of the local design team and the construction team.

The project manager was not responsible to the owner for the co-ordination and control of the overseas consultants' work. The project manager was only responsible to the owner for the co-ordination and control of the local consultants' work.

The relationships between the owner, overseas consultants, local consultants, the project manager, and the contractors created a significant problem in this project. The problem arose when the contractors proposed design changes (e.g. the replacement of imported materials with local materials). The decision to approve, or not, the changes took a long time and was held to be the main cause of delays on the project. The contractor proposed such alterations to the project manager. The project manager would then ask the local consultant to analyse the proposal. If the changes were still within the specification requirements, the project manager would consider the alterations and forward them to the owner. The owner gave its final approval after discussing the proposal with the overseas consultants.

This co-ordination problem might not have happened if the owner realised the situation earlier and changed the approach by giving more responsibility to the project manager to manage the interface of the consultants.

The parties involved in the project did not change their approach when they found out that the relationships between parties might cause problems that lead to project delays.

Performance

The period for completion of Contractor 1's work was set at 36 weeks with the total estimated cost of approximately US$13M. A number of problems in finishing the work (e.g. the replacement of imported materials with local materials) extended the completion date. It had been anticipated that there would be delays. However, quality levels were achieved.

Although the delays had been anticipated, by giving the project manager more responsibility to co-ordinate and control both the overseas consultants' and local consultants' work, the delays might have been shortened.

Multiple contracts versus single contract

This project used multiple contracts. However, a single contract was considered – the main contractor would subdivide the work into several work packages which would be tendered to nominated subcontractors.

EXERCISE
1. Once a procurement system is in place, how difficult is it to change it to another?
2. Approval delays were a problem on this project. Suggest a satisfactory way of reducing such delays.
3. Assess the arguments given for using either one or two contractors.

5.2.6 CASE STUDY – AIRPORT EXTENSION, CAMBODIA

Introduction

An airport extension project was undertaken in Cambodia. The aviation authority engaged a consultant company to prepare the concept design, detail design, and contract documentation and to supervise the project.

After the design stage, a work breakdown structure of the project was developed and it divided the project into:
A = Civil works and building construction.
B = Road building, sewage and stormwater drainage pipe installation and furnishing for the new airport terminals.
C = Services (power, communication, gas and water reticulation).

Three successful bidders were engaged and lump sum contracts (no extras, variations, ...) entered into with the authority. Contractor A was in charge of civil works and building construction. Contractor B was in charge of road building, sewage and stormwater pipe installation and furnishing, and Contractor C was in charge of services.

All three contractors acted as head contractors and subdivided their work to subcontractors. Contractor B entered into lump sum contracts (no extras, variations, ...) with two subcontractors. Subcontractor B1 was responsible for sewage and stormwater pipe installation and Subcontractor B2 was responsible for road building.

However, one month after the end of the tender evaluation process, aggregates and asphaltic concrete increased 80% on their normal price due to some unforeseeable political reasons. As a consequence of that, both subcontractors suffered increased costs of materials. Subcontractor B2 was likely to end up with a loss and was trying to discharge the contract. Subcontractor B2 did not continue with the work, while Subcontractor B1 decided to continue in order to maintain the company's reputation.

As part of their contracts, both Subcontractors B1 and B2 had to complete a portion of the roadworks within three months. This portion of the roadworks was a milestone for the whole airport extension project. Because Subcontractor B2 was trying to discharge the contract, that portion of the roadworks was one week behind schedule. Subcontractor B1 had to wait for Subcontractor B2 to undertake excavations. Finally Subcontractor B2 returned to work and continued construction when it found out that it did not have a valid reason to discharge the contract.

For a portion of the roadworks in progress, two design errors were identified by Contractor B. The designated location of a pump station did not match the actual location, and there was a conflict in levels between the sewer and stormwater pipes. Also, there was conflict during construction due to the poor work breakdown. For example, there were pipelines, designed to pass through sewer manholes, that were included in the contracts of Contractors A and C rather than with Contractor B. Thus, sometimes there were two or three subcontractors working in the same area and doing similar work, and this caused a waste in manpower and equipment. Fortunately, the design and supervision were under the responsibility of the same consultant company and the errors were rectified by redesign within a week. The new design adjusted the pump station twenty metres closer to a stormwater tank and also adjusted the sewer pipe one metre higher than the original design; this benefited Subcontractor B1 because there were less earthworks and there were savings in stormwater pipe material. However, these design errors still interrupted the work to both Subcontractors B1 and B2. This portion of the road building was finished two weeks behind the contracted date.

The relationships between the owner, consultant, head contractors and subcontractors are shown in Figure 5.8.

This delivery method was adopted because the detailed design was completed in advance of tendering and the allocation of funding. Some disadvantages could be eliminated by adopting other methods and are discussed below.

EXERCISE
1. The owner could have allowed extras, variations, ... (due to changing costs of material and labour, and changing scope of work) in the lump sum contract, especially to allow for unreasonable inflation in the costs of material and labour which might affect the performance of contrac-

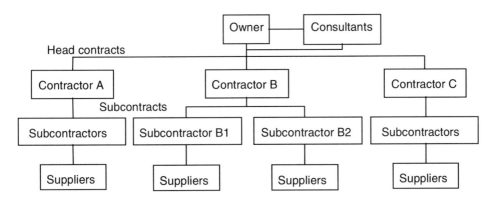

Figure 5.8. Delivery method, airport extension project.

tors and cause potential conflicts between the owner and contractors. The contractors could have tried to reduce their costs whenever it was possible and to minimise the quality aspects. Discuss.

2. The owner could have engaged a construction manager to provide a service for the construction phase. The work breakdown may have been thought through more clearly by an experienced construction manager. If an experienced construction manager had been engaged, then it could have designed the interfaces of the contracts, thereby reducing the potential for conflicts between the owner and contractors, but also reducing the costs for both the owner and contractors. Discuss.

3. In this project there were multiple contractors involved and this required considerable coordination. If a construction manager had been engaged on this project, then it may have managed the coordination of contractors better. In addition, involving a construction manager avoids the problem where the design and supervision are done by the same firm – it is not easy for the consultants to realise or admit their mistakes. Discuss.

4. A well written specification can be achieved by being viewed by people other than the author. There is a suggestion that not only the owner but also the consultants and contractors should check the specification to eliminate errors. In addition, designers and others need to check the designs before the tender process starts, in order to eliminate errors and omissions, and avoid delays during the construction process. Discuss.

5. The owner could have adopted an incentive and penalty scheme instead of only adopting penalties. The owner could have used bonuses and penalties for completion of the project ahead of or behind schedule, respectively. For example, the owner could have adopted an incentive scheme that may have prompted Subcontractors B1 and B2 to increase their productivity, whereby the bonus would be used as extra profit to counter the increased costs of materials. Incentive and penalty schemes can be useful in time-constrained projects. Discuss.

6. The lump sum contract is favoured by owners because the risks are transferred to the contractor. However, local practice is for the owner to also carry some risks, because consultant firms tend to overestimate part of the bill of quantities. The reason for overestimation is that, if the bill of quantities was underestimated, this would cause the contractor to suffer loss. The contractor would then criticise the consultant firm for underestimating the bill of quantities, and the consultant firm might lose future work. Comment on this local practice.

7. In this case study a redesign was required in the fixed price contract (without rise and fall). The consultant considered the contractor's situation and the redesign tended to favour the contractors. Comment on this local practice.

8. How do you deal with the possible situation where material prices might increase by 80% – consider from the contractor's and subcontractor's viewpoints?
9. What better work breakdown structure could have been chosen to eliminate overlap between the contractors?

5.2.7 CASE STUDY – B&T DELIVERY, TOLL ROAD, INDONESIA

B&T is an acronym for Build and Transfer. This delivery method was used on the Taman toll road project – a road of about 25 km in mountainous area, with bridges and tunnels. The total scope of work involved soil investigation, alignment, detail design, construction, and even land acquisition. The project owner was a state-owned company that managed toll roads. The B&T project was awarded to a state-owned contractor company.

This road was a part of the overall road network. There were about ten similar projects being developed at the same time. The owner wished to engage companies on a BOT contract basis for all projects. This strategy could not be applied to the Taman toll road because the feasibility study did not promise a reasonable profit, and so investors were not too interested in taking the project as a BOT deal.

The discussions proceeded by negotiation and without competitive tendering. The contractor submitted a brief offer based upon the concept design provided in the feasibility study and a toll road design standard. The evaluation and negotiation were based upon the unit price of other similar projects. After a clarification and negotiation phase, the owner and the contractor signed a schedule of rates contract. At the beginning of the contract period, the contract amount was based upon item unit prices. At the end of land acquisition, road alignment and soil investigation were expected to be completed. The contractor and the owner then negotiated the total price of the project and signed a lump sum contract.

All funds were arranged by the contractor. For the cost of money, the contractor received a fixed rate of interest on total cost. The contractor would be paid for the 15-year period after completion. During these 15 years the contractor would receive a monthly payment of an amount 80% of the total income collected from the toll road. A sum of 20% was to be kept by the owner for the operation costs of toll road activities. At the fifteenth year, the balance of the contract amount will be paid off, clearing all debt.

There were no bonds in this contract. Guarantees were based on trust and faith, between the two state-owned companies.

EXERCISE
1. How can the public be sure that their money will be used efficiently in this contractual arrangement?

5.2.8 CASE STUDY – LUMP SUM CONTRACT, DISCHARGE, CIVIL WORK,
INDONESIA

Outline

A civil work project was executed in Indonesia, as initial work for the construction of a geothermal power station of capacity 2×65 MW. The owner of the project was a consortium of local and overseas players, while the main contractor was a large US contractor. A major part of the civil work was earthworks and gabion slope protection, which was carried out by an Indonesian construction company, under a subcontract amounting to approximately US$5M. This case study is written from the viewpoint of the subcontractor.

Two issues

Two main contractual issues were of interest and are discussed in this case study. The first issue deals with (sub) *contract type*, particularly concerning the lump sum price, and the second one, concerning *discharge* of the (sub)contract. As the subcontract was on a lump sum basis, the scope of work was supposed to be well-defined and all documentation reasonably complete. The geological investigation data were incomplete. Due to this incomplete geological information, the subcontractor suffered extra costs for equipment to carry out cut-and-fill work in the main power station area, and associated extra time.

By reason of unsatisfactory performance of the subcontractor (denied by the subcontractor), the contractor discharged the subcontract at the 70% progress stage. The main reasons the contractor argued were insufficient safety measures and slow progress. In the view of the subcontractor, these reasons were not supported by factual data.

EXERCISE
1. On this project, the subcontractor accepted a lump sum work offer without receiving complete information about the geotechnical investigation, particularly for the platform area of the main power plant. Only brief information regarding this was provided at the tender stage.

   Unfortunately, when the subcontractor carried out cut-and-fill work in the main power plant area, an unexpected geological condition was encountered. At a certain level, the area could not be cleared by the 'normal' equipment combination (standard bulldozer and dump trucks) because the soil condition was too muddy. The subcontractor had to supply a different bulldozer (for swampy terrain) and double-wheel steering dump trucks. This situation forced extra costs on the subcontractor, and delays (over that scheduled and allowed for in the bid).

   Soon after, the subcontractor submitted a claim to the contractor for cost compensation based on the insufficient geological information supplied by them, leading the subcontractor to suffer extra cost and extra time. By returning this claim, the contractor denied the claim, and pointed out a clause in the general conditions of the subcontract, saying that the subcontractor is deemed to have examined all conditions on the job site included its surrounding area at the tendering stage.

   There was no price adjustment or provisional sum clause in this subcontract.

   What is the subcontractor's position? Does the subcontractor have any recourse to the contractor's decision?
2. What is the contractor's position?
3. How does the lump sum contract type influence your thinking?

Subject index

adjustments 27
agency arrangement 90, 141, 148
alliances 92
attendance 28

bill of quantities 30, 32
bill of quantities contracts 29, 38
BLO 86, 154
BO 86, 154
BOLT 86, 154
bonus 55
BOO 86, 154
BOOT 86, 130, 154
BOT 86, 154
bundling 176

CM 142
commercial development 86, 154
competitiveness 14
concessional methods 86, 154
concessions 154
construct only 86, 101
construction management 86, 90, 132, 136, 137, 141
construction manager 141
construction project management 86, 90
contract (payment) type 1, 25, 62
contract approach 6
contractor 3
convertible contracts 26
cost plus 41
cost plus contracts 40
cost reimbursement contracts 40
cost savings 14

D&C 86, 107, 113

day labour 5, 42
daywork 5, 42
DCM 86, 130
DD&C 86, 107
delays 27
delivery method 1, 83, 86, 92, 93, 179
delivery method selection 83, 84, 85
design novation 86, 121
design-and-construct on two levels 117
design-and-construct(ion) 86, 107, 113, 122, 129, 135, 137
design-build 86, 107
design-construct 86, 107
design-construct-maintain 86, 115, 129
design-develop(ment)-con-
struct(ion) 86,107
design-document-con-
struct(ion) 86, 107
design-manage 86, 107
detail design-and-construction 86, 107, 111
direct control 5
direct labour 6
disincentive 25, 55
divided contract 175
do-and-charge 40

early involvement 88
engineering, procurement, construction and management 86, 141
EPCM 86, 141
escalation 27
extras 27

fashion 180
fast-track 92, 164

fee 41, 43, 45, 50, 56
firm price 28
fixed fee 46
fixed fee with bonus/penalty 48
fixed percentage fee 45
fixed price contracts 25, 27, 29
fragmentation 181

gainshare/painshare 60
gender comment 4
guaranteed maximum cost 50
guaranteed maximum price (GMP) contracts 40, 50

in-house approach 5
incentive 55
incentive contracts 40
incentives/disincentives 55
independent contractors 142
integrated contract 86
issue matrix 94, 177

latent site conditions 28
lump sum contracts 26, 29, 30
maintenance contracts 61
management contracting 86, 90
management contracts 45, 90
managing contractor 86, 107, 120, 149
measurement contracts 29
multiple contracts 175, 177

negotiated contracts 41
nominated subcontracting 122
non-agency arrangement 90, 148
novation 121

outsourcing 5, 6, 8, 12, 87

owner 3
owner's brief 109
owner's representative 3
owner-builder 86, 147
owner-contractor relationship
    3

package deal 107
packaging 175
partnering 92
payment 103, 108, 113, 123,
    124, 132, 143
PCM 142
penalty 25, 55
performance contract 107
period contract 92
piece rates 29
PM 133
PPM 133
prime cost 41
prime cost (PC, pc)
    items/sums 28
prime cost contracts 25, 40
private sector work 9, 12

procurement 1, 2
project management 86, 90,
    131, 135
project manager 131
provisional items/sums 27, 28
public sector work 9

quantum meruit 42

range of incentive effective-
    ness 56
reimbursable costs 43
relationship contracting 92
RIE 56
rise and fall 27
risk 26, 94, 97, 124, 137, 159,
    180

schedule of rates 29
schedule of rates contracts 26,
    29, 33
selection, contractor and con-
    sultant 9
separate contractors 142

sequential contracts 178
serial contracts 178
single contract 175, 176
specialist contractors 142
staged contracts 178
staging 178
subcontracting 102

target estimate/cost/price con-
    tracts 48
terminology 3, 25, 28, 29, 41,
    55, 92, 133
trade contractors 142
traditional method 86, 101,
    114, 131, 135, 137, 145,
    148
turn key 86, 107

unbalancing 35
unit price contracts 29

variable (sliding) percentage
    fee 47
variations 27, 110